D1289550

Exposing the
Magic of Design

Oxford Series in Human–Technology Interaction

Series Editor
Alex Kirlik, University of Illinois at Urbana-Champaign and the Beckman Institute

Being There Together: Social Interaction in Shared Virtual Environments
Ralph Schroeder

Human-tech: Ethical and Scientific Foundations
Kim Vicente
Edited by Alex Kirlik

Adaptive Perspectives on Human–Technology Interaction: Methods and Models for Cognitive Engineering and Human–Computer Interaction
Edited by Alex Kirlik

Computers, Phones, and the Internet: Domesticating Information Technology
Edited by Robert Kraut, Malcolm Brynin, and Sara Kiesler

Neuroergonomics: The Brain at Work
Edited by Raja Parasuraman and Matthew Rizzo

Attention: From Theory to Practice
Edited by Arthur F. Kramer, Douglas Wiegmann, and Alex Kirlik

Information Foraging Theory: Adaptive Interaction with Information
Peter Pirolli

Exposing the Magic of Design

A Practitioner's Guide
to the Methods
and Theory of Synthesis

Jon Kolko

With contributions from:
Beth Johnson and
Gianna Marzilli Ericson, *Design Continuum*
Paul Gould, *MAYA Design*
Colleen Murray, *Jump Associates*
Hugh Dubberly, *Dubberly Design Office*
Lauren Serota, *frog design*
Rachel Hinman, *Nokia*

OXFORD
UNIVERSITY PRESS

OXFORD
UNIVERSITY PRESS

Oxford University Press, Inc., publishes works that further
Oxford University's objective of excellence
in research, scholarship, and education.

Oxford New York
Auckland   Cape Town   Dar es Salaam   Hong Kong   Karachi
Kuala Lumpur   Madrid   Melbourne   Mexico City   Nairobi
New Delhi   Shanghai   Taipei   Toronto

With offices in
Argentina   Austria   Brazil   Chile   Czech Republic   France   Greece
Guatemala   Hungary   Italy   Japan   Poland   Portugal   Singapore
South Korea   Switzerland   Thailand   Turkey   Ukraine   Vietnam

Published by Oxford University Press, Inc.
198 Madison Avenue, New York, New York 10016
www.oup.com

Library of Congress Cataloging-in-Publication data

Kolko, Jon.
   Exposing the magic of design : a practitioner's guide to the methods and
theory of synthesis / By Jon Kolko.
        p. cm. — (Oxford series in human-technology interaction)
      ISBN: 978-0-19-974433-6 (hardback : alk. paper)
    1.   Industrial design—Psychological aspects 2.   Product design.
I. Title.
   TS171.4.K648   2011
   745.2—dc22   2010036267

To Jess, just for being there . . .

Contents

Introduction

Designers continually describe their profession as a way of organizing complexity or finding clarity in an overwhelming amount of data. Jeff Veen, founder of leading design consultancy Adaptive Path, has noted that "Good designers can create normalcy out of chaos" (Veen, 2000). Jim Wicks, vice president and director of Motorola's Consumer Experience Design Group gives the name "synthesis" to this ability to create normalcy. As he explains, design always includes ". . . synthesis—synthesis of market needs, technology trends, and business needs" (Wicks, 2006). During synthesis, designers attempt "to organize, manipulate, prune and filter gathered data into a cohesive structure for information building" (Kolko, 2007b). *Synthesis reveals a cohesion and sense of continuity. Synthesis indicates a push toward organization, reduction, and clarity.*

Yet despite the acknowledged importance of this phase of the design process, synthesis appears magical when encountered in professional practice. This sense of magic is both good and bad. The idea of designers as magicians is an intriguing metaphor, because their work is mysterious and the output can be phenomenal and tremendously emotional. Yet for those who value logical and linear thinking—both engineers and business owners, respectfully—design synthesis is a frustrating part of product development. Because the act of synthesis—the revelation of clarity—is frequently performed privately ("in the head" or "on scratch paper"), the *outcome* is all that is observed, and only after the designer has explicitly begun the form-making process. Unlike other aspects of the design process (such as drawing, which even a naïve or detached audience can observe and generally grasp), synthesis is often a more insular activity, one that is harder to understand or even impossible to see. A designer may follow a user-centered discovery process to immerse himself in a particular subject or discipline, and he will then "incubate" that material.

After a period of reflection, he produces a tangible artifact to visually represent the reflection. Synthesis conducted in private produces no visible connection between the input and the output. Often, not even the designer himself can articulate the exact value of his design insights. He must create a plausible rationale after the fact, or the client must trust him. More often than not, the client simply rejects the insight as being "blue sky," "unfounded," or simply too risky.

For example, a designer developing a new digital device might study the use of digital devices in the workplace. Typically, the designer would observe four or five users as they work. The designer would question the users about their jobs and record their responses. The designer might also take screen shots or photographs of the tools being used and probe for details about each tool. Then, in the privacy of her studio, the designer would attempt to make sense of what she has learned. Her goal would be to find relationships or themes in the research data and to uncover hidden meaning in the observed behavior that applies to the design task at hand.

The user-research sessions would have produced pages of verbal transcript, hundreds of pictures, and dozens of artifact examples. Because of the complexity of thinking of so much data at once, the designer would turn to a large sheet of paper and a blank wall to "map it all out." Several hours later, the paper would be covered with what, to the uninitiated, appears to be a mess. Yet the designer has made substantial progress, and the mess actually represents the deep and meaningful sensemaking that drives innovation. The designer will have identified themes in the data, and she will better understand the problem she is trying to solve. The designer will have discovered "the whole," as described by Daniel Fallman, a professor at Umeå University: "Fieldwork, theory and evaluation data provide systematic input to this process, but do not by themselves provide the necessary whole. For the latter, there is only design" (Fallman, 2003).

A Lack of Method in Practice Leads to Problems

Methods of contextual research are increasingly common in the development of new products, systems, and services, and they are performed in more progressive companies to observe users in order to learn from them. Researchers may use primary research tools, such as ethnography, quantitative surveys, or more traditional forms of market research, to gather data from their customers. Bodies of secondary research describe how the brain works, how people solve complicated and multifaceted problems, and how to apply creative thinking to policy or social science. And most design literature emphasizes learning by doing, explaining that there must be an explicit translation phase of research in order to make sense out of it.

Yet rarely do practicing designers have sufficient time to perform design synthesis in a rigorous, substantial manner. Anecdotal evidence indicates that even the most advanced and well-known corporations touch only tangentially on issues of synthesis, interpretation, and the rigorous process of translation—of making sense of the gathered research—during their normal work activities. And while academic research has described how people make sense of complexity, this research is published in exclusive scientific journals that rarely find their way into design consultancies or large corporations. Even when they do, they often reference other papers and require a long trail backwards to become understandable and actionable.

Compounding the problem, the designers who actually build the systems, services, products, and artifacts that we use on a daily basis self-characterize as being overworked and too busy to delve into the complicated literature of tangential disciplines. They require something more immediate and approachable if they are to integrate new ideas into their design process. They do not lack the intellect to understand the complexity of academic research; they simply lack the time to read it at all.

Thus, practicing designers are not using the strong set of grounding theories for design synthesis. To an observer (commonly a client), the physical output, the themes, and the design ideas produced seem arbitrary, or *magically derived*. The artifacts developed by the designer are messy, usually drawn in the midst of deep and reflective thinking. They are sketches drawn in Sharpie, incomplete sentences, and crude diagrams without adequate captions or descriptions. If the beginning state (the research data) is compared to the end state (the design idea), it is not immediately evident how one led to the other. In fact, it can be argued that *the more innovative the output,* the more difficult it is to identify how the idea was developed. Yet the incubation period described earlier can be well structured, and the things that occur during that period are both repeatable and comprehensible. It is only the lack of understandable documentation, or the decision to not share that documentation, that creates the sense of magic.

Clients may well desire magic, because it suggests that they have spent their money well (after all, they have hired magicians or shamans!). But the notion that design synthesis is magical and therefore difficult to formalize has led to four very large problems that plague the industries of designed artifacts:

1. *Clients do not see the relationship between design research and design ideas; therefore, they entirely discount the value of design research and design synthesis.* Because synthesis is frequently relegated to an informal step in the overall process, it is practiced implicitly and little physical, obvious evidence is produced to support it. After several design projects that include implicit design synthesis, a client may proclaim that he does not see the value in a discovery phase for future design activities. Of course he is right: He did not *see* anything of value, so he assumed the phase to be a waste of resources.

2. *Design consultancies do not plan for, assign resources to, or appropriately bill for synthesis activities, and so design synthesis happens casually or not at all.* If no formal period of time is allotted for design synthesis methods and no formal deliverables are associated with these methods, the designer receives a strong message: Synthesize on your own time, or do not synthesize at all.

3. *When synthesis actually occurs, it is perceived as messy, and so it gains a negative reputation as an overly complicated or difficult part of the process.* The output of

design synthesis is frequently incomplete or intangible—the value of the output is not immediately evident, because the results are "half baked." Synthesis often results in high-level themes and paradigms that help shape future design activities, but these conceptual elements may be seen in retrospect as too abstract as to justify the time and resources spent.

4. *Novice designers do not know how to accomplish synthesis, so they flounder through this portion of the design process.* These methods are rarely formalized in design education, and experienced designers have taught themselves to achieve synthesis in their heads in an informal, reflective manner. A beginning designer has few places to turn while attempting to solve design's intellectual problems.

These problems are roadblocks to innovation, and they illustrate a deep disconnect between the core process of insight development and the billed process of product development. Yet synthesis methods have been continually referenced as critical in sensemaking and organization, and in drawing important connections between elements that appear to be unrelated. These are the keys for relating research to design. Synthesis methods are the ways in which ethnographic insights are mapped to new, innovative, appropriate, or compelling ideas.

These principles and methods are teachable, repeatable, and understandable. They are creative activities that actively generate intellectual value and that are unique to the discipline of design. Most important, when applied and formalized, these activities are billable and immensely useful in the development of novel, useful, and appropriate designs.

The Goals of This Text

There are three goals for this text. The first goal is *to present a theory of design synthesis* in a simple and concise manner. This theory is based on academic research and discourse, but it is presented in a way that is clear and valuable to a practicing design manager, designer, or design researcher. This theory of design synthesis can then be used to substantiate single methods of synthesis.

The second goal is to *offer a rationalization of why design synthesis is important*, both in a general sense ("Why should I care about this at all?") as well as in a more immediate sense ("Why should I care about this right now?").

The final goal is *to present a set of actionable, learnable methods for design synthesis* that can be applied to any design problem. Practicing industrial designers, interaction designers, interface designers, and designers of other disciplines can use these methods to make sense of complicated design problems and to move seamlessly from various forms of research to design. The methods can add a systematic sense of rigor to an otherwise subjective, often introspective process.

This text is an introduction to design synthesis for practicing designers and business owners. The text presents a framework of synthesis, borrowing heavily from research related to human behavior. Within this framework, the text also presents a series of design techniques that can be immediately applied to both big and small design problems. Finally, more advanced applications of synthesis illustrate the potential for complicated problem solving.

Many designers struggle with *innovation expectations*. They are asked by their company or client to develop new, interesting, and marketable designs on demand, much as a dog might be expected to roll over. If you are in this situation, you may wish to skip directly to the Methods section, where you will find a number of techniques you can apply almost instantly to stimulate your creativity.

Like many designers, you might also question the intellectual roots of your work when asked to give clients a cohesive theoretical substantiation for what automatically appears during design. You may wish to read only the Theory section, because these chapters describe the psychological frameworks that underlie much of the work you already do.

Designers, both new and experienced, will likely benefit from reading the whole text in order, because it offers both a theoretical foundation as well as multiple repeatable design methods for generating meaning and taming complexity. You will also benefit from the whole if you are a business owner, a product manager, or a subject-matter expert in a large corporation that finds itself in the anxiety-ridden role of developing new products, systems, and services.

The Immediacy of This Text

For increasingly complex issues of sustainability, finance, culture, and technology, businesses are searching for a form of problem solving that can deal with the unprecedented levels of ambiguity, chaos, and data. Businesses that need a way to stay competitive in a global economy have been urged to reject standard forms of marketing and product development. The popular media has disparaged traditional "linear thinking" as inadequate for the global economic crisis. Yet little has been offered as an alternative. It is not enough to demand more innovation without providing the tools to succeed. Design synthesis is a way of thinking about complicated, multifaceted problems of this scale with a repeatable degree of success.

Exposing the
Magic of Design

Section One
What Is Synthesis?

Chapter 1
A Theory of Synthesis

Synthesis in design involves the combination of two complicated entities: the designer and the design problem. The theoretical grounding presented in this chapter emphasizes the unique qualities of the designer (her experience, expertise, and the complexity of her design and personal experiences) and the unique qualities of the designer's frame of the design problem (the inherent constraints and her mental model of the problem). The designer and the design problem engage in a dance of process, creativity, and often, conflict.

A consideration of this theoretical grounding of synthesis offers two main benefits to the reflective designer. First, this theoretical grounding acknowledges the complexity of the designer, and it begins to hint at what makes a "good" designer "good." Through the designer's experience, he has been able to develop knowledge that extends beyond the domain of a specific design sector (mobile, Web, pharmaceutical, retail) and into the actual process of design. With a fair degree of autonomy, an experienced designer can therefore understand, rationalize, and better frame a given design problem. The designer develops unique constraints that are not part of the original client brief and understands how these constraints directly contribute to his ability to solve the given problem. Secondly, the theoretical grounding describes a foundation upon which the "magic" of design occurs. This is the cognitive rationale for why design happens. It explains why designers are able to take incomplete data, manipulate it in various ways, and invent things that are relevant, innovative, or appropriate.

This chapter explains the fundamentals of design intellectualism. Because synthesis is a process of making meaning, it is not unique to design. In fact, for years, researchers from other disciplines have been investigating how people make meaning. Computer scientists have examined how people solve problems, often with the goal of creating problem-solving systems. The larger umbrella for this research—artificial intelligence—is well known, but many of the theories that ground it are hidden in arcane journals and rarely discussed in relation to aesthetics or product development. Similarly, university researchers of communications and information sciences have attempted to craft a unified theory of how people transmit information. Although this research could have immediate resonance in our work, too often it is hidden in university libraries and mired in complicated medical prose.

Understanding How People Solve Problems

The late Herb Simon, one of the most prolific social and computer scientists of the 20th century, spent much of his professional career trying to discover *how people decide*. Although his work initially focused on pragmatic administrative and economic theories, he soon gravitated toward understanding how people solve problems, because problem solving is ultimately a process of decision making.

Discussion of problem solving would seem to be reserved for complicated fields like physics or chemistry, but in reality, we all solve countless problems each day. From the "problem" of choosing what to wear in the morning, to the "problem" of finding an e-mail buried in a messy inbox, our lives are filled with issues to be solved. Often these problems are hierarchical. For example, trying to locate an old e-mail may simply be a way to dig up information needed to solve a larger problem. Consider the following:

1. [The large problem] I need to find the best and most appropriate people in my organization to assign to a specific project and get them involved with the project.

    a. [An embedded problem] I need to find an old e-mail that describes the availability and skill set of Jeff Smith.

i. [An embedded problem] I do not find the e-mail when I do a text search for Jeff's name.

Simon's interest in problem solving stemmed from another interest: artificial intelligence. He investigated how people conduct complicated activities, with the goal of modeling and eventually reproducing these actions using a computer. Computers are good at replicating mundane or procedural tasks. For example, if I want a computer to flag every e-mail that contains the words "Jeff Smith," I can create a simple procedural rule and the computer will follow my instructions.

This type of problem is "well structured." Simon defined this type of problem as fulfilling the following criteria (Simon, 1973):

1. It includes criteria to test the solution and a repeatable process for applying the criteria. In the earlier e-mail example, the criterion is finding a specific word in the body of the e-mail.

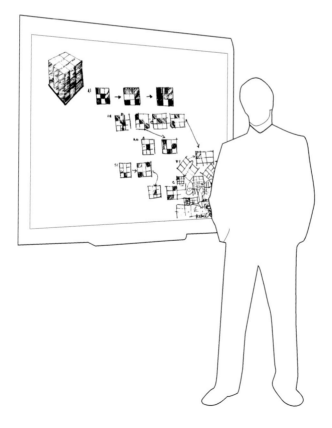

Figure 1.1
Hierarchical problems.

2. We can identify the initial problem state (e-mails exist; do they have the phrase "Jeff Smith" in them?); the goal state (we know that each e-mail either does or does not contain the phrase); and the interim states (we search through each e-mail, word by word, until we have searched the whole e-mail).

3. We can identify "legal moves"—that is, the steps we take to solve the problem that fall within the logical constraints of the system.

4. We can identify any knowledge the problem solver has about the problem.

5. The "legal moves" required to solve the problem reflect the laws of nature; for example, one cannot expect the computer to read minds.

6. The "legal moves" required to solve the problem require only practical levels of effort.

If a problem is not well structured, Simon described it as "ill structured." Design problems related to innovation and creativity—the types of problems facing designers of products, software, and services—are almost always ill structured, because they do not fulfill the aforementioned criteria. One cannot ask a computer to "come up with the next big thing" or "develop the new innovative product." Computers have difficulty with even seemingly small ill structured problems, like the initial problem mentioned earlier ("I need to find the best and most appropriate people in my organization to assign to a specific project and get them involved with the project"). There are a few things that make it ill structured:

1. What does the "best and most appropriate talent possible" mean in this particular project context? What are the criteria for deeming something "best"? What kind of talent?

2. What does it mean to "appropriately assign a person to a project"? Is this a match based on experience, skills, availability, or personality?

3. What criteria can we use to test the solution? Can we compare and contrast what would happen if we assigned Jeff Smith, as compared to Helen Jones, to the project?

Although this problem is ill defined and hard for a computer to solve, it is quite solvable by a person. People confront this type of problem successfully in their jobs every day.

Consider what happens as our scenario plays out:

I need to find the best and most appropriate people in my organization to assign to a specific project and get them involved with the project. It is a project that requires some deep experience with mobile messaging, and I recall that Jeff Smith worked on something similar to this in the past. The project was successful, and I am pretty sure it was because of Jeff's contribution. If I remember correctly, Jeff has a really positive attitude. He can work with everyone, and although he is a leader, he allows others to feel as though they have come up with the big ideas.

I search through my e-mail, in an effort to locate an old e-mail that described when Jeff Smith would be available for a new project and that also provided details about his skill set. I remember that his manager sent this a few months ago.

I cannot find the e-mail when I do a simple text search for Jeff's name because there are a number of Jeff Smiths in the company. After several more attempts, I give up. I pick up the phone and call his manager directly. We have a brief discussion, and he tells me that Jeff is able to tackle a mobile-messaging program, and he is available and eager to participate in this type of project again. It is confirmed, then: Jeff will start ASAP.

In this scenario, I have made substantial progress, yet I have made a number of leaps that a computer would have a hard time emulating.

1. I am "pretty sure" that Jeff was essential to the previous project's success, but I have no objective evidence to support this. It is a memory, and it might be wrong.

2. I decide that Jeff is the best talent possible with little formal or repeatable evaluation of other choices. In fact, I have focused in on Jeff very quickly—there may be someone much better, but I do not even consider other candidates.

3. I cannot remember much about the e-mail, except that it came from his manager.

4. I subvert the entire e-mail search by using a different communications mechanism: the phone.

These leaps represent typical interpretative (and subjective, but not necessarily haphazard) steps seen during problem solving of ill-defined problems: acting on an informed hunch, making a judgment, using partial information, and, after working within constraints to no avail, completely breaking out of these constraints.

Acting on an Informed Hunch

We often make decisions in the "blink of an eye"—using our intuition and acting on an informed hunch (Gladwell, 2007). To do this, we use tacit knowledge—knowledge built up over time, through experience—concerning the problem. In the earlier example, *the project was successful, and I am pretty sure it was because of Jeff's contribution.* I have gathered evidence, through hallway conversations and through observational evidence, to draw a fragile link between Jeff's contributions and the project's success. Additionally, I am willing to act on that link. The problem solving depends on my ability to commit without a complete picture of the problem. Without a complete picture, the information I have will have to *satisfice*. The idea of *satisficing*, and the word itself, are central to Simon's theory of bounded rationality, which implies that people will make rational decisions given the limited information they have at a given time.

Making a Judgment

I judge Jeff to be the best possible talent for the position, yet I never consider other candidates. This is a subjective, evaluative activity, and I might be completely wrong. Instead, I have synthesized various "moving parts" to the problem, over time, and somewhat autotelically. My judgment—given the above bounded rationality—allows the process to continue and for an active decision to be made. To some degree, once I have made this decision (perhaps without even being aware that there was a decision to make), it is incredibly difficult to undo the decision. Even with an attempt at objectivity, I will likely ignore evidence that discounts my decision and embrace evidence that supports it.

Using Partial or Incomplete Information

I cannot identify which "Jeff Smith" is the correct one or remember enough about Jeff to successfully search through the data I have gathered (i.e., my e-mail inbox). Yet somehow I have associated all of the subjective, evaluative ideas about Jeff's abilities to an incomplete idea of him.

This particular association describes how priming can create a sense of spreading activation, leading quickly from one idea to another. The words "Jeff Smith," combined with my contextual cues (resourcing, job, e-mail search), allow me to "think around" the problem and eventually land on the proper memory of Jeff's manager. I can think of all of the related aspects of this particular Jeff: the last time I saw him, the horned-rimmed glasses he wears, and his red hair. All of these traits help me hone in on the particular idea of Jeff in memory, and this honing ability is critical to problem solving.

Understanding, and Breaking, Constraints

Perhaps the most important part of problem solving is the ability, and tendency, for successful decision makers to work around or completely ignore formal or technical constraints. In this case, the "problem constraints" included e-mail and a search mechanism. Yet a human element holds not only the answers to the question of "Who is Jeff?" but also additional subjective, qualitative—and important—data about Jeff's ability and availability for the project. Eventually, I subvert the system by calling Jeff's manager. I may have broken a rule, if there was a formal project allocation process of assigning people to projects. But I solved the problem.

These problem-solving techniques—acting on a hunch, making a judgment, using partial information, and breaking constraints—point to the foundation of a theory of synthesis. They are *human,* not *logical,* techniques. The distinction is important, because it points to the complexities of ill-structured problem solving. It is difficult to generate a single model of the human approach to solving ill-structured problems, so it is equally difficult to produce tools to *help* people as they solve these types of problems. Yet the strategic and innovative portions of business are ill defined and chaotic. Rarely does a design team have the complete picture and a strong, organized set of procedures for translating research into meaningful design insights. Problems of innovation in business are almost always ill structured, just like the one discussed in this chapter.

Chapter 2
Sensemaking, Frames, Models, and Patterns

Cognitive psychologists Robert R. Hoffman, Gary Klein, and Brian M. Moon define sensemaking as "a motivated, continuous effort to understand connections (which can be among people, places and events) in order to anticipate their trajectories and act effectively" (Klein, Moon, & Hoffman, 2006a). This process occurs over time as one or more people try to connect disparate pieces of data. Their definition is related to one of organizational theory; in a large organization, various people may hold different pieces of data that are all critical to the success of a product or project. Someone may go through a sensemaking process by collecting these pieces of data, meeting with all of the other players, and bringing the data to a central place where it can be formed and manipulated.

This definition builds on communication theorist Brenda Dervin's own theory, which implies that we learn when we make meaning ourselves. Rather than absorbing facts as bits of predigested elements (the way we might learn Spanish by repeating vocabulary words to a tape), we make sense of complexity by *doing* things (the way we might learn Spanish by visiting Spain for an extended period of time).

Consider the student who is miserably failing a required chemistry class. He does not understand the formulas being presented or even why he should *care* to understand them. The class is boring and predictable: The professor comes in, writes on the chalkboard, and the students take notes. Class after class, nothing stands out, and the student has trouble remembering anything at all.

Then, one day, the professor shifts from a lecture style to an experimental style. As class starts, the professor pours two liquids together and a giant fireball shoots out of the beaker. The student is intrigued. After overcoming his surprise, the student's mind is filled with thoughts of other fires he has seen: cigarettes, candles, gas stove burners, campfires. Some of these thoughts are triggered automatically, whereas others are purposefully recalled through experience cognition, such as "That reminds me of when John lit the napkin at that restaurant. I can't believe he was so clumsy; we almost got kicked out!"

In this example, the student has made connections between the experiment and his life. He has integrated the class experience into his world of knowledge and made associations between them. Because he has made these connections, he is more likely to remember the class and make it meaningful. If the professor is further able to hold this student's attention through the presentation of the formulas and can tie this representational information to the visual process of fire just demonstrated, the student can forge powerful connections between a symbolic illustration of a reaction and the experiential and emotional example of it.

This is sensemaking in action, albeit in a highly reductive example. It is learning in a way that draws from the unique, subjective, and rich experiences of the student. He is forming associations and connections between some formulaic and objective data, some vivid and sensory experiences in the classroom, and the rich, personal knowledge he has gathered through his life experiences.

The point is that in design, one should not try to escape one's past, emotions, or upbringing in an attempt to be more "objective" because these elements are central to making sense of a complicated problem. Two designers may approach the same problem in the same way and follow the same methods and steps, yet they may end up with very different solutions. This difference points to the unique aspects of the designers themselves—their "style," or "design sensibility"—which is the collective and additive whole of their lives. Design is not entirely subjective, nor is it entirely objective, but it is both at different moments.

An understanding of sensemaking is important because synthesis taps deeply into the ability of a designer to judge, through a highly subjective frame, the design problem she is solving. An awareness of the frame itself can lead to this form of judgment.

The Role of Perspective in Framing Situations

A frame is an active perspective that both describes and perceptually changes a given situation. As described by Klein, Moon, and Hoffman, "even though frames define what counts as data, they themselves actually shape the data (for example, a house fire will be perceived differently by the homeowner, the fire fighters, and the arson investigator" (Klein, Moon, & Hoffman, 2006b). A frame is, simplistically, a point of view. Commonly, the point of view has little objectivity, so it is often deemed "irrelevant" or "biasing." In this way, a point of view may differ from moment to moment and can be thought of as a short-term perspective. But a frame, shaped over the aggregation of thoughts and experiences, is a larger view of the world and situations that occur in it. Like a point of view, a frame too will change, but over the long term rather than the short term.

We use frames to make sense of situations. Consider the house fire described earlier. The homeowner, obviously distraught, may react to the event in one of the following ways:

- "I just can't get a break. My life is a series of unfortunate events, and this is just another example of my bad luck."

- "Thank goodness we escaped successfully. It's only stuff—it can all be replaced—it's just important that we are still alive."

- "My life is over. All of my important things were in that house, and I don't know how I'm going to ever be able to get my life back together again."

Each of these statements is shaped by a lifetime of experiences, and in turn, this frame will continue to shape further action and behavior. If the homeowner views her life as over because her objects were destroyed, she is approaching the situation from a materialistic standpoint. The homeowner illustrates, through this statement, that objects and artifacts have a direct connection to her self-image. We can infer with some degree of accuracy how this type of person might react to another situation, because the "materialistic frame" transfers.

In design, framing can be thought of as the designer's perspective when approaching the problem (both conceptually and pragmatically). The frame itself applies a set of exterior, subjective constraints to the design problem; it is built on the types of experiences referenced during sensemaking.

Take the real-world example of designers who have been tasked with building software for use with a consumer wireless router, which will bring Internet access into a home and then distribute it. The client has provided a set of practical constraints: The software must be easy to use and must provide access to all of the functionality provided by the router. The client may even have quantified the constraints in a requirements spreadsheet that says things, such as "The user will have the ability to connect to the Internet" or "The user will have the ability to enable port forwarding."

The design team can frame this problem from a conceptual point of view in any number of ways. Consider these *high-level frames*:

- *Ease of use.* The user should never encounter confusing things or technical jargon.

- *Power.* The software should afford complete control over the robust feature set of the router, so the user should be able to manage even the most nuanced setting on the router.

- *Pleasure.* The act of using the router software should be pleasing and emotionally fulfilling. The user should always feel a deep and emotional response to the various routing activities.

In fact, each of these frames was suggested by various designers on this project. Not surprisingly, the first frame was suggested by a designer with an interest in usability engineering, the second by a more technical designer, and the third by a designer who specialized in visual interface creation.

Exposing the Magic of Design

All of the frames add extra constraints to those supplied by the client, yet these extra constraints are not a burden. In fact, they allow the designers to move forward with their work, as they funnel the realm of "all possible choices" into a much smaller and workable set of "appropriate choices."

Framing manifests itself at a much more detailed level, too. Consider the following conversation that occurred much later in the router-software design process:

> Designer: "When the user clicks on the image of the router, the context menu should animate outwards and to the right, as though the user caused the menu to appear."

> Developer: "But if it animates to the right, it will cover whatever content is there. Shouldn't the user be able to see all of the content on the screen?"

> Designer: "Maybe, but it's more important that the user feels like he or she caused the menu to appear—it's important that we show a sense of causality on mouse-click."

> Developer: "But it's going to cover things up. That seems completely inefficient."

Because the designer's and developer's frames are different, they result in different practical design considerations. The designer is pushing for a more aesthetically compelling animation that reinforces causality. The developer is more interested in presenting an uninterrupted set of content. In effect, the designer is drawing on a frame of temporal aesthetics, whereas the developer has framed the problem in a context of utility.

The Importance of Models in Sensemaking

Sensemaking and framing can be enhanced and supported through externalization and through representations. Common to all methods of synthesis presented later in this text is a "sense of getting it out" to identify and forge connections. This is an attempt to make obvious the sensemaking conditions described earlier. Emphasis is placed on finding relationships and patterns between elements and forcing an external view of things. In all of the methods, it is less important to be "accurate" and more important to give some tangible form to the ideas, thoughts, and reflections. Once externalized, the ideas become "real." They become something that can be discussed, defined, embraced, or rejected by any number of people, and the ideas become part of a larger process of synthesis. Essentially, sensemaking is an internal, personal process, whereas synthesis can be a collaborative, external process.

In his landmark text *Notes on the Synthesis of Form*, Christopher Alexander describes two major techniques for learning to build new artifacts. The first is a master–apprentice model, where a new designer learns by being constantly surrounded by a particular craft and ultimately acquires skills through practice. The second learning technique occurs through externalization, as the new designer "...tries, in some degree, to make the rules explicit" (Alexander, 1964). In a design problem, the rules are often implicit in the problem and considered design constraints.

Constraints might be obvious, such as "you have this much money to spend to solve the problem," but frequently, a host of much more subtle, and important, rules are deeply embedded in the subject matter. A designer makes these explicit through a model. Alexander explains that "what does make design a problem in real cases is that we are trying to make a diagram for forces whose field we do not understand," that designers often try to *solve* a design problem while simultaneously trying to *understand* the design problem. Modeling delineates between problem finding and problem solving, acting as a form of problem understanding. Ultimately, this is the role of synthesis: to create a stage for problem understanding.

A model is subjective and interpretative. As interaction designer Hugh Dubberly explains, "Models help bridge the gap between observing and making, between research communities and design communities. Models are especially important in interaction and service design" (Dubberly, 2009). He further describes a model as a conjecture acting as a reasoned way of proposing how things might be. The conjecture is most useful when approached with an open mind for change. A designer will find that a model offers the most value when considered as an indication of thinking at a moment in time, not as a finished artifact. In fact, a comparison of an early model and a subsequent model can offer deep insights into the thought process and rationale for various design decisions.

Mental Models as a Specific Type of Cognitive Representation

A mental model can be thought of as an inaccurate yet helpful representation of how something in the world works. Although the Dubberly and Alexander models described in the previous section are often tangible (diagrammatic and thus can be drawn on a piece of paper), mental models, as the name implies, are stored in our memory. These models act as the frame, usually when a designer approaches a complicated system, in which to act and react.

Psychologist Kenneth Craik, one of the first to consider the concept of mental models, describes how a mental model acts as a posit of "what if"—a point of departure, around which multiple ideas can be considered. As he writes, if the person "carries a 'small-scale model' of external reality and of his own possible actions within his head, he is able to try out various alternatives, conclude which is [the] best of them, react to future situations before they arise . . ." (Craik, 1967). A mental model is actually a way of understanding how things work and exist in three dimensions and in the fourth dimension of time, even if we cannot "picture" the thing in our mind.

Imagine a door lock. This is actually a fairly complex mechanical system, and so it should be, because we depend on it to keep important places and things private, safe, and secure. Yet this complexity is obscured for most people, who have created a mental model of how these items work. When most people consider a lock mechanism (if they consider it at all), they might think of the shape of the key, inverted. That is, "If I place a key in a lock, there's a similarly shaped hole that makes room for the key. If the key and the hole match, the door opens, and in I go." This is technically and logically inaccurate, as a series of tumblers, wedges, and other intricacies make a lock and key work, yet the shorthand model allows us to go about our lives without tending to the minutiae of devices and systems. We can imagine why a key does not work and construct—again, incorrectly—a view of how to fix it.

An engineer typically creates a mental model framed in pragmatism and logic. He or she knows how things *actually* function, so he or she views the world from that perspective. In the same way, an environmentalist approaches the world framed in sustainability, and a visual artist views the world from an aesthetic perspective. Our ability to frame a problem, and the mental models we create to cope with complexity, are obviously intertwined and heavily dependent on our lives, careers, knowledge, and abilities.

Philip Johnson-Laird references these types of models as critical to our ability to reason. As he describes, "We use perception, the meanings of words and sentences, the significance of the propositions that they express, and our knowledge. Indeed, we use everything we've got to think of possibilities, and we represent each possibility in a mental model of the world . . . a model isn't an image, but the abstract structure that underlies images and that represents contents that we cannot visualize" (Johnson-Laird, 2009).

As a more germane example, consider a designer working through the synthesis phase of design. To develop an innovative system for growing vegetables, the designer has learned a great deal about lettuce through contextual research with farmers. Now he is sketching his mental model of how vegetables grow. If that model is legitimate, he understands the relationship among sunlight, soil nutrients, and temperature, so he sketches something realistic (see Fig. 2.1).

Continuing from this sketch, the designer can develop a number of ideas for enhancing plant growth. A standard representation of a mental model may lead to a standard set of outcomes.

But suppose the designer purposefully embraces the following incorrect and incomplete mental model of causality: "The farmers put the seeds in the ground, and there's some sort of reaction with the dirt. The better the seed is planted— the more dirt that can touch the edges of the seed—the better the reaction. When the farmer waters the seed, it's a way of getting more dirt to touch the seed. There seems to be an

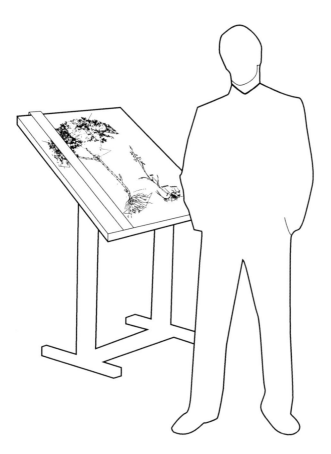

Figure 2.1
Mental Models form during the
process of design.

opportunity here. I can design a system where dirt is packed tightly around the seed and is constantly shifted to replace the old dirt with new."

The mental model seems plausible, and the designer is able to temporarily suspend disbelief to see what happens. The designer has honed in on causality (albeit logically inaccurate causality) in his sketch: A lot of dirt touching a seed leads to a big lettuce plant. Ideation can now occur again. The designer can develop a number of new ideas for getting as much dirt as possible to touch the seed (flattened seeds, a dirt mover, etc.).

These ideas might offer inaccurate growing techniques, and they simply might not work. But by temporarily embracing an obviously incorrect mental model, the designer has developed a new opportunity and space for design ideas that otherwise would not have been considered. And these new considerations can be then reintroduced into the more mundane, "appropriate" design path.

The Nature of Patterns on Our Experiences

As designers become more seasoned, they build up a level of experience and expertise that lets them act as though the output is "intuitive." They seem able to solve design problems effortlessly. In fact, design never becomes effortless, but the process of design becomes increasingly fluid and amorphous as the designer becomes more capable, confident, and reflective.

This is the role of experience in a field where every problem is unique. Experience—the number and scope of design problems the designer has previously encountered—plays a critical role in developing design synthesis expertise. Rich design experience affords the following five core benefits:

1. An experienced designer can generalize a design problem to a "typical" design problem. He or she is able to find patterns in the design problems she comes across. Once a pattern has been implicitly identified, the designer is able to generalize both the problem-solving process and past solutions to this type of problem.

2. An experienced designer can anticipate how the problem will unfold and will react to various design activities. This means that fruitless efforts are minimized. By anticipating how they will play out, the designer can pursue ineffective solutions with less frequency and in less depth.

3. An experienced designer can anticipate how external factors, such as client or stakeholder feedback and changing market conditions, will affect a design decision. By doing so, the designer can either mitigate these factors or build time into the schedule to respond effectively to them.

4. An experienced designer is faster at sketching, modeling, and other methods that are critical for both understanding and illustrating design synthesis to other stakeholders. So he can move quickly from the "mess" of research to the beginnings of a design solution.

5. An experienced designer knows, implicitly, what to do next rather than wasting time and effort fretting over it. At a fundamental level, an experienced designer will sketch the aforementioned models, without external prompting.

All of these benefits create the perception of intuition, as a designer's activities then appear effortless and immediate. A great deal of this is due to the highly subjective identification of patterns: the ability to find repeated methods, processes, and knowledge in a large set of data, and to abstract away the specific differences.

Patterns contribute to our ability to make decisions, and our language skills help us to recognize large patterns. These patterns help designers make progress on complicated design problems. Design patterns are both trends and common approaches to particular design challenges. These patterns can be large or small, general or specific, but they always illustrate something that has appeared in the world more than a few times. As a general pattern, consumers are becoming savvier with small-screen interfaces, due to their interactions with ATMs and cell phones. As a more specific pattern, many products are positioning media in a carousel metaphor; users can "spin through" their media, as is the case with Apple's coverflow or HP's TouchSmart. Both types of patterns represent something repeated enough in popular design culture to have affected the way people view and understand a product.

A pattern can be applied and connected with other design elements to reach a new conclusion or change a design. In that case, it provokes a new way of thinking. A pattern identified in research acts as a design guideline or a constraint to help shepherd further ideation. In both cases, the designer must be able to match incomplete data in various contexts to find commonalities, yet without strong or cohesive guidelines. Simply, with few rules to define what is and is not considered a pattern in design, the designer is left to subjectively decide when enough commonality exists to call something "patterned."

Experience contributes to pattern generation, which, in turn, drives successful design work and a more fluid process.

Chapter 3
Abductive Reasoning

Synthesis is an abductive sensemaking process. Abduction can be thought of as the "step of adopting a hypothesis as being suggested by the facts . . . a form of inference" (Peirce, 1998a). Roger Martin, dean of the Rotman School of Management at the University of Toronto, has described this as the "logic of what might be"—abduction is the *argument to the best explanation*. It is the hypothesis that makes the most sense given observed phenomenon or data and based on prior experience. Abduction is a logical way of considering inference or "best guess" leaps (Martin, 2009). To better understand abduction, it is necessary to understand deduction and induction.

A valid deductive argument is one that logically guarantees the truth of its conclusion, if the premises that are presented are true. This form of logic is traditionally taught in mathematics courses and manifested in logic proofs:

A is B.

All Bs are Cs.

A is, deductively, C.

This form of logic is self-contained, and any argument that uses deduction cannot offer any *new findings* in the conclusions—the findings are presented in the premises that hold the argument to begin with. That is, A, B, and C all exist in the presented premises.

As another example, I can say that "Google is a Web site" (A is B). I can also say that "All Web sites can only be accessed when the computer is connected to the Internet" (All Bs are Cs). Deductively, then, "Google can only be accessed when the computer is connected to the Internet" (A is, deductively, C).

If both initial statements are true, the conclusion is true, too.

An inductive argument is one that offers sound evidence that something might be true based on structured experience. This form of logic is traditionally associated with scientific inquiry:

> *Each time I do A under the same conditions, B occurs.*

> *Inductively, the next time I do A under these conditions, B will occur.*

Subsequent experiences may prove this wrong, and thus an inductive argument is one where the premises do not guarantee the truth of their conclusions. Like deduction, induction cannot offer any "new findings" contained within the logic of the argument.

As an example, I note that "each time I visit Google when I'm not connected to the Internet, it doesn't work" (Each time I do A under the same conditions, B occurs). Inductively, I infer that "the next time I visit Google when I'm not connected to the Internet, it won't work" (The next time I do A under these conditions, B will occur).

Abduction is where the magic happens.

Consider the following example:

> *If I do A, B occurs.*

> *I've done something like A before, but the circumstances weren't exactly the same.*

> *I've seen something like B before, but the circumstances weren't exactly the same.*

> *I'm able to abduct that C is the reason B is occurring.*

Unlike deduction or induction, abduction allows for the creation of new knowledge and insight. C is introduced as a best guess for why B is occurring, yet C is not part of the original set of premises. And unlike deduction, but similar to induction, *the conclusions from an abductive argument might turn out to be false, even if the premises are true.*

As an example, I note that "if I visit Google when I'm not connected to the Internet, I would expect that it wouldn't work. But one time, it does! I see a list of only a few results, but they aren't very thorough" (If I do A, B occurs). I've visited Google before when I wasn't connected to the Internet (something like A), but I wasn't sitting on a subway car (the circumstances weren't exactly the same). I've gotten Google results before (something like B), but they were more thorough and more numerous (the circumstances weren't exactly the same). I'm able to abduct that the subway is somehow acting as an intranet service to present me with incomplete data access, giving me "enough," but not "everything" (I'm able to abduct that C is the reason B is occurring).

Design synthesis is fundamentally a way to apply abductive logic within the confines of a design problem (Coyne, 1988). The various constraints of the problem begin to act as logical premises, and the designer's work and life experiences, and her ease and flexibility with logical leaps based on inconclusive or incomplete data, begin to shape the abduction. Abduction acts as intuition and is directly aided and assisted by experience of any design or cultural patterns. As described by Peirce, "The abductive suggestion comes to us like a flash. It is an act of *insight*, although extremely fallible insight. It is true that the different elements of the hypothesis were in our minds before; but it is the idea of putting together what we had never before dreamed of putting together which flashes the new suggestion before our contemplation" (Peirce, 1988b).

The earlier example—Google on the subway, working offline—offers a new idea of a way of providing subway riders access to small bits of data without installing expensive cabling or a lot of equipment. It might not be a great or technically feasible idea. But it is a *new* and *insightful* idea, one that came from a process of abductive and hypothesis-driven exploration, not from the original constraints.

Johnson-Laird has indicated that in the context of generative and creative problem solving, the insight is not developed in a "flash" at all. Instead it comes through a four-step process whose goal appears instant when achieved but is really reached through a formal and methodical (but internal) thought process. This process looks like this:

1. The current problem-solving strategy (almost always inductive or deductive in nature) fails to yield a solution, given the existing constraints.

2. There is a tacit consideration of new constraints.

3. The constraints are relaxed (or changed) in a new way, thus broadening the problem space and allowing for further consideration.

4. Many changes in constraints lead nowhere, but, with perseverance, a change may be made that leads at once to a solution of the problem. (Johnson-Laird, 2005)

These four steps can be made more obvious through the design methods found in later sections of this text. Step number three is critical: Constraints, things that have been established as either implicit or explicit boundary conditions, are *changed*. In the context of a business problem, this might mean that budget is increased, delivery platform is changed, or features are added or removed.

Both Peirce and Johnson-Laird agree that abductive reasoning is related to insight and creative problem solving, in that this form of reasoning *creates something new*. Researchers Hideaki Takeda, Akira Tsumaya, and Tetsuo Tomiyama have formalized this idea in their description of knowledge integration in the synthesis process:

> ... the aim of synthesis is to create objects having necessary characteristics. In this case, it is not required that characteristics are universal and minimum, rather they should not be. In order to capture human desire for objects, characteristics should be as rich as possible to represent various desires. Thus requirements for knowledge for synthesis are not universality and minimality but rather individuality and diversity. The last statement indicates that the assumption underlying the traditional logical approach is not appropriate because it is to capture our world with minimum and universal axioms. (Takeda, Tsumaya, & Tomiyama, 2007)

Synthesis, then, is about creating a quantity of newness, where each new idea is individually unique. Roger Martin (2009) agrees, describing a knowledge funnel that exists as the environment for business innovation. This funnel describes the space in which decisions of "newness" occur in a business context. In this space, design ideas move from mystery, to heuristic, to algorithm, and finally to code. In the context of design and business, Martin has found that designers prefer to live in the "mystery" phase, producing as many new and diverse ideas to address the mystery as possible. Everything begins as a mystery, and through a process of questioning, such as "Why do apples fall and birds don't," we gain a first level of understanding. We develop a heuristic, or a general way of thinking about things. We formalize this into a predictive-based rule—an algorithm—and finally, we arrive at a place where we can describe this rule in enough detail to automate fully and to capitalize on the rule and the outcome of the rule-based process.

By contrast, business owners focus on this rule-based process or code and find the most excitement in moving toward reliability and repeatability. At the beginning of the funnel, the goal is to develop ideas that embrace, as Tomiyama et al. describe, "individuality and diversity." At the end of the funnel, the goal is to have identified a single or core idea and formalized it in such a way that a machine can duplicate it. The goal is to develop something "minimal and universal" (see Fig. 3.1).

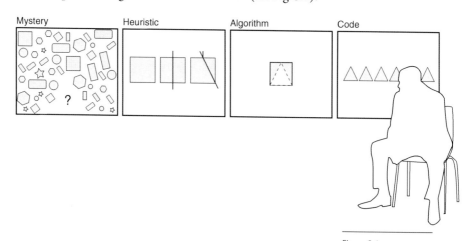

Figure 3.1
The Knowledge Funnel.

That creates a gap between the mysterious dreaming that designers appreciate and the proof business owners demand. Design cannot prove the solutions to the mysteries because solutions are only demonstrable and testable, not provable. The idea of proof relies on and insists on both inductive and deductive reasoning, which wipes out new and innovative thinking. The form of abductive thinking described earlier by Peirce, and echoed by Johnson-Laird, lives at the beginning of the knowledge funnel. It acts as a means of understanding the mystery and forming heuristics of ideas. This form of abductive thinking is design synthesis.

Section Two
Design Synthesis in a Business Context

Chapter 4
The Value of Synthesis in Driving Innovation

As the word *innovation* has crept into the vocabularies of executives, so too has the word *design*. The search for the keys to innovation has made increasingly clear that both "design thinking" and ethnography play a critical role in the larger context of the design process. In this context, businesses are increasingly realizing that not just quantitative research but also qualitative research combined with creative thinking can lead to new and interesting ideas for products, services, and systems.

Observational research is the type of qualitative design research often performed at the beginning of the design process. This research involves observing real people, in their environments of work and play, as they go about performing a task, achieving a goal, or having an experience. Pragmatic design research grew out of information technology and software design, as researchers observed people using complex computer systems in an effort to increase the usability of these systems. Other qualitative research methods have latched onto the successes of the social sciences in understanding culture. Ethnography involves immersion in the culture in which people spend their time, a method that is much less goal directed and might simply offer a topic statement—the "culture of entertainment," for example—as a starting point.

Design research is different from marketing research. The goal in design research is to find inspiration for design, whereas the goal in marketing research is to predict the behavior of a larger group. Unfortunately, large businesses nearly always lose this critical distinction in the similar terminology and approaches. Consider the core similarities and distinctions listed in Table 4.1.

Table 4.1 Core Research Similarities and Distinctions

Design Research	Marketing Research
Focuses on people	Focuses on people
Can be qualitative or quantitative	Can be qualitative or quantitative
Borrows from the social and behavioral sciences	Borrows from the social and behavioral sciences
Attempts to *understand culture*. Looks at the styles, words, tools, and workarounds people use in an effort to inspire design.	**Attempts to *predict behavior*. Looks at what people say they would do, or what they actually do, in an effort to predict what they would do in a new situation**
***Celebrates* the unique and peculiar. The rare or obscure in observations can lead to a new or interesting design idea.**	***Avoids* the unique and peculiar. The goal is to understand mass responses; outliers are frequently ignored.**
Avoiding bias is *irrelevant*. The goal is not to be objective but instead to be rigorous.	**Avoiding bias is *critical*. The statistical analyses of data require an objective point of view.**

To witness the ambiguity about research in the context of a business problem, consider the real-world example that follows:

A design team is writing a proposal for a $200,000, 12-week program to develop innovative digital-file storage. A third of the time will be spent conducting research, and the team decides to observe how eight people work and document that research through transcription. From this research, the team will produce a list of key insights and use those insights to generate preliminary design ideas.

When the designer presents the proposal to the client, the following dialogue occurs:

Prospective Client: "So, I see here that 4 weeks—or about $65,000—is going to research. We've already done a lot of research. We have a great document that we can give you. It's about 60 pages long with lots of charts and graphs. It describes the segments we are targeting and has responses to a great survey we conducted with thousands of people. So let's take this section out completely."

Creative Director: "Although your document sounds very interesting, the type of research we do is different. We will look for some of the nuances in the work environment, learn about the culture of digital storage in the workplace, and really get to know how people think about digital file storage."

Prospective Client: "Your proposal shows you're only going to look at eight people, though. We've done a questionnaire with thousands of online respondents, so we feel pretty confident that we know what people want."

Creative Director: "Understood, but it's a really critical part of our process . . ."

Prospective Client: "I would really like to see this proposal rewritten without that $65,000 in it."

Both sides are right, but they are right about different things. The client's questionnaire data—if it was collected properly—can indicate a set of statistically relevant predictions about what a larger population *might do*. But this data will not describe *what to make, how to make it*, or what the interactions and experiences *should feel like*. Because both sets of activities are called research, the client is quick to dismiss this seemingly duplicative work as a waste of time and money. Then the designer's difficult job becomes to educate and evangelize the unique role of design research. Designers may succeed in educating the client by showing samples from previous projects (and highlighting how research was conducted) and will sometimes show diagrammatic representations of how the research data will be transformed directly into the final product.

An unfortunate side effect of this dialogue is that unless the education succeeds, the designer will likely still conduct the research but will not charge for it. This will diminish the value of design research on that particular project and, over time, design research as a whole.

Jan Chipchase, a design researcher at frog design, commonly conducts research without a particular project or product in mind (and frequently without the challenges of the evangelizing described earlier). Chipchase was asked by frog to travel the world, observing people and their communication behavior. He was not looking at traditional usability, such as the "ease of use" of specific products or the number of taps it takes to dial a number. Nor was he focused on marketing metrics or the search for "innovation" or "insight." Instead, Chipchase worked to understand how communication fits into culture, be it augmented by technology like a cell phone or facilitated by a human mechanism like dance, speech, or writing.

Whether the research is pragmatic—as in looking for usability enhancements—or conceptual—as it was for Chipchase—the philosophy of design research is the same: to learn from people and to emphasize people, rather than technology or business. As an example, consider a design-research program focused on understanding the social relationships between teens and mobility. A researcher could approach this problem from three immersion perspectives by immersing herself in the following:

1. A group of teens who frequently travel a great deal or commute to school and then to their jobs. The researcher would try to understand the way teens keep in touch over distances, and she would learn about the language and the feelings about staying in touch over distances.

2. The technology used by teens, by looking at their various computers, phones, and other technological devices. The researcher would try to understand the pros and cons of various existing tools, and she would learn about the attitudes toward these devices, the most and least frequently used features, and the qualitative feelings about the various tools.

3. The business of mobile communication and networking, by looking at the services and capabilities of leading companies such as Facebook or AT&T. The researcher would learn about pricing models, tiered offerings, branded services, and the other packages.

In the first perspective, the researcher will learn about *potential for the future state*. She will see problems as opportunities, and findings will emphasize behavior. Behavioral opportunities may exist in the following areas:

- Technology (the devices the teens own)

- Style (the clothing the teens wear)

- Identity (the language the teens use)

- Infrastructure (the transportation methods the teens use)

This perspective is broad. The researcher can learn about the topic from a holistic sense, looking both directly at the problem of "social relationships and mobility," as well as *around* the problem.

In the second perspective, the researcher will learn about *the current state, from a realistic standpoint.* The word *realistic* implies that the artifacts being investigated have conformed to the realities of production. Their constraints are usually pragmatic ("At the date of production, a smaller phone could not fit all of these electronics. No matter how much we want it to be smaller, it cannot be"). The researcher will gain a great deal of knowledge around the following areas:

- Usability of existing software, hardware, and services

- Frequency of use of existing software, hardware, and services

- Emotional resonance of existing software, hardware, and services

This is useful information, particularly when attempting to fix and refine an existing solution with a "follow-on" release. The perspective is narrow because the researcher focuses on particular devices and the software functionality afforded by those devices.

In the third perspective, the researcher will learn implicitly about *the current state, from a subjective standpoint.* A particular service is offered for free or for a cost because a business decided it would be so ("We can offer this service for free in order to build brand equity, which is worth more to us than the amount we could have made from charging for the service itself"). The researcher will be able to gain a great deal of knowledge around the following areas:

- Pricing, feature, and service structures other companies have deemed important

- The way various artifacts, services, and systems have been positioned in the marketplace

- The elements that have become ubiquitous to a particular business context

This is useful information when attempting to package and sell an already designed product, service, or system. The perspective is narrow. The researcher focuses on particular pricing structures, service offerings, and capability models from various competitors.

Each of these research methods focuses on a different aspect of an artifact, yet only the first—focusing on human behavior—emphasizes *opportunity* and *potential*. A focus on technology or product is destined to be constrained by realism (e.g., What can we do, given current abilities?). A focus on business or market is destined to be constrained by precedent (e.g., What are others doing, and how are they doing it?). In this way, design research that focuses on human behavior in a broad sense—not on a particular object or service—is the most effective at discovering data for innovation.

Figure 4.1
Various Perspectives Feeding Design.

Exposing the Magic of Design

What Is Innovation?

Innovation has been used so liberally to define an entire profession that one is hard pressed to find a definition of the word itself. It is best used as either a simple qualifier that can be used to describe one facet of design—newness—or as a title for a robust and entirely different field.

Researchers Craig Vogel, Jonathan Cagan, and Peter Boatwright use an action-oriented context for their definition: Innovation "extends beyond invention of new technology and includes a thoughtful and insightful application, delivery, extension, or recombination of existing technologies . . . the key is that an innovation is a valued leap from the viewpoint of consumers whether or not it is incremental from the producer's standpoint" (Vogel, Cagen, & Boatwright, 2005). This definition puts the consumer at the center of the "innovation universe," and so it seems logical to then emphasize the value of design research. It is important to note, however, two major problems with viewing design research as the "keys to innovation."

First, an innovative product is not simply new; it must be new *and successful* in the marketplace. This means that a new idea that fails—for example, Betamax or the Apple Newton—cannot be considered truly innovative. This is not simply a matter of semantics. Design-research methods will help a product team find the newness, but they will not help bring that newness to market. For this, traditional forms of design, engineering, supply-chain management, quality, and other production techniques are necessary, as are traditional forms of marketing, advertising, and distribution.

Second, and more important, design research presents only an opportunity, but it does not lead directly to the new idea or innovative concept. Most businesses understand how to conduct research (either quantitative or qualitative, and often marketing driven but occasionally user centered). Fewer businesses understand how to design something. But it is the rarest of companies that can continually drive innovative design practices and actually tie the newness to the research that was conducted. The link between initial investigative efforts and subsequent creative efforts is rarely, if ever, emphasized.

Design Synthesis Links Innovation Research and Design

Design synthesis is the link between the type of behavioral research described earlier—the potential for the future state—and the creation of something new. It is the most critical part of the creative process of design. Yet many designers rely only on their own limited experiences in approaching design synthesis.

In the generative stages of a design problem, designers often turn to pencil sketching on paper to think through the various nuances. For example, to visualize the appropriate form of a new touch-based cell phone, an industrial designer will sketch in three dimensions and in orthographic (or plan) view, often laying ideas on top of one another and switching between a stylistic approach to a more pragmatic, component-based investigation (looking at the actual elements that might need to be contained within the phone, such as a screen, a keypad, and so forth). At this ideation stage, the most high-level design problems have been defined, so the designer is *problem solving*. That is, the designer knows what he is creating—a phone, and not a toaster or a printer—and he knows the general constraints of the object (it has a certain-size touch screen and requires a certain-size battery to power it, and so forth).

But consider the previous stage, in which the high-level design problems are defined or identified. Why *isn't* the designer creating a toaster, for example? It may be that the company in question has a high degree of competency and history in creating mobile phones. Or the company may have developed a new technological approach to building low-cost touch screens, so it is trying to find new applications for it. Or it may be that the company has identified, through research, a new opportunity for producing a touch-based phone.

Where do these discussions happen, and who has them? Typically, these types of considerations are made by directors of marketing and technology. These organizational structures control a big budget, which they (often independently) assign to whichever projects and programs they deem to be most strategic. Once they have made the decision, a product team is assembled. Eventually the product "trickles down" to the designer, who then begins to sketch what the item might look like.

But with the recent popularity of the phrases "design thinking" and "innovation," designers have been asked to participate in these strategic conversations. Designers are increasingly expected to discuss not just how to solve a problem but also which problems to consider solving. They are increasingly pressured to speak with clarity about product launches, strategic product road mapping, competitive marketplace trends, short- and long-term revenue opportunities, partnerships and sponsorships, and other issues related to the business of design.

This presents a great opportunity for designers to move from a tactical role to a strategic role, where they are valued not only for their ability to produce but also for their ability to think and analyze. Yet even at these more fundamental levels of a design problem, there is an implicit expectation that the designer is *designing*—producing things that are visual and tangible, that trigger additional discussion and that evoke emotive responses. Essentially, if a designer is to enter the boardroom, she is expected to bring something unique to the boardroom discussions.

What are these *unique things*? What does the designer do or make while attempting to find and understand problems at a strategic level?

Design synthesis generally describes this aspect of design, where the designer is not yet solving a problem but is still doing, and making, in an attempt to understand complexity. Synthesis is an intellectual approach to creativity, and it can offer a rationalization for repeated business success and a set of tools for moving from research to specific and actionable design ideas. Because synthesis is tied to logical processes of managing complexity, it can be communicated throughout an organization and used to substantiate the seemingly "magical" world of design and design thinking.

A designer attempting to produce an innovative design will conduct research focusing on the experiential, emotional, and personal aspects of culture. This research will describe an opportunity—design research acts as problem *finding*. The research findings may be captured in PowerPoint presentations or described on a whiteboard. Either way, the research has allowed the design team to gather data within a constrained problem space.

Design is that act of problem *solving*—of appropriating formal qualities into a new design idea that fulfills the stated criteria and adds value to the human condition. Design synthesis, then, will translate the opportunity into specific design criteria, or a set of elements that must be present to afford a cohesive and concrete design. The synthesis will describe the solution; design synthesis is the process of problem *understanding*. Although data gesture toward an opportunity, data are frequently thick and convoluted, overwhelming and incomplete. The data alone lack contextualized meaning, and so it is difficult to decode data in their "raw" state. Synthesis is a sensemaking process that helps the designer move from data to information, and from information to knowledge.

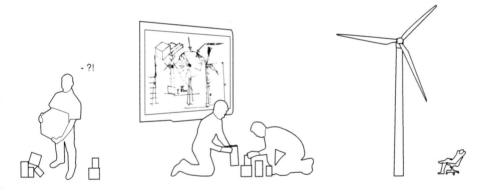

Figure 4.2
Problem Finding, Understanding, and Solving.

Exposing the Magic of Design

Chapter 5
The Culture of Synthesis

Simply put, synthesis is a spark. It is the ability for the human mind to grasp multiple, often incongruent and even competing ideas, and to manipulate them—at once, and in parallel—into something amazing. Synthesis allows for multiple hypotheses, ideas, themes, patterns, or trends to be mapped and diagrammed, and consumed and explored. It is a process of judging, yet it celebrates the cultural nuances that form the judgment. Synthesis is elusive, yet is not magical—it only feels that way because its output is new, exciting, untamed, and full of potential.

Creating the spark of synthesis requires a constant evolution of ideas through an informed trial and error, and the creation of form where there was none. But this presents a conundrum: When design is found in the context of business, the oversized external constraints quickly overwhelm the space in which creativity can occur. These external constraints include billability, deadlines, or resourcing. While some designers can push past these, even incorporate them in their understanding of a design problem, others allow these to chip away at their creativity. The lack of inclusion or appropriate time, or the "just get it done" attitude, forces the good to become the quick, with sloppy results that lack a sense of honesty, passion, or completeness.

In the past few years, a proliferation of companies, Web sites, and articles have claimed to improve a company's ability to be creative or, even better, to innovate. These resources commonly recommend brainstorming, carrying a notebook, and prolifically reading, yet none of these resources ever describe how to gain the motivation to be creative. It is as though they assume that the right mixture of activities leads directly to creative output. It is commonly understood that the passion and fire of creativity have to come from within, and yet the burdens of beginning, and finding the energy and patience to see creative ideas through to fruition, are often what can stop the creative individual from executing at all. In fact, if we break the problem of creativity into two parts, we can begin to see a very obvious disconnect between *having creative ideas* and *creating things*.

As business owners follow the buzz of innovation as a meme to replace quality, they subscribe to literature that recommends a number of pragmatic steps to increase the creative mindset of employees. These include the following:

- Building a culture that is forgiving of mistakes

- Encouraging fun, humor, and a playful environment

- Supporting business decisions that have ethnographic rationalization

Yet good ideas come through a number of sources, all of which are related to humanity. Whether the source is ethnography, popular culture, or even the humanitarian realness of the world around, new insights into old problems come through observation and synthesis, through conversation and discourse.

But within these points lies a subtle trap. The designer needs to explore the world and accumulate patterns, drawing connections in real life and empathizing with the richness of humanity. She needs to try and fail, to play, joke, and jest. Yet while the designer explores, learns, and plays, she is not *designing* in the generally understood definition of the word. Design requires the production of something with some sort of form (even considered loosely, as in the design of systems, and services, and political systems). Simply "having a good time," "watching people," or "thinking really hard" does not constitute this form giving, and although potentially generative, the lack of tangible output hinders the practicing designer.

This illustrates the dilemma of the professional designer: A designer is judged on his ability to produce, but one cannot observe production during idle reflection on culture. Those who are not producing—or, in the case of a design consultancy, billing their necessary hours, are a detriment to the productivity of the studio and are ultimately let go. Yet an underlying and obvious demand is that the designers who bill their time are designing something with both purpose and appropriateness.

If design requires reflection and absorption of culture, both the studio-consultancy system and the corporate design system are set up to encourage a failure on both accounts for designers who are not taught or allowed to challenge constraints, act in a playful manner, experience flow, and use visual thinking as their primary mechanism of thought and communication. This is the culture of synthesis, and it is a culture that can be fostered and nurtured. The sections that follow describe these qualities that are instrumental to driving innovation and supporting design synthesis.

Challenging Constraints and Questioning Purpose

Although the word *constraint* has a negative tone, constraints are one of the central tools for managing an otherwise overwhelming design activity. Designer Charles Eames constantly referred to constraints as the most important aspect of creating a successful design. As he describes, "... here is one of the few effective keys to the design problem—the ability of the designer to recognize as many of the constraints as possible—his willingness and enthusiasm for working within these constraints— the constraints of price, of size, of strength, balance, of surface, of time, etc.; each problem has its own peculiar list" (Neuhart & Neuhart, 1989). Eames is describing the qualities that contain a design problem, that mark its beginning and ending, and that illustrate to what extent the designer can affect change.

Essentially, Eames is indicating a core distinction between design and art. For all practical purposes, art has no external constraints. The artist selects the constraints, ignoring those that she feels are inappropriate and embracing those that help tell a story, make meaning, or create a particular aesthetic. But the designer cannot ignore or embrace the borders and guidelines on a whim; she must meet a deadline, present a particular message, or solve a given problem.

When applied during the spark of synthesis, constraints are made explicit. This subtle and simple point is often daunting to a novice designer because the most important constraints that are to be made explicit have yet to be defined. They do not come from the project brief or even the manager. In most complicated design problems, the client provides a core set of constraints, often called "requirements" and presented in a document in the project brief. As Henrik Gedenryd describes in his thesis on the cognition of design thinking, ". . . constraints are regarded as given to the designer, as part of the requirements specification, before design begins. Moreover, they make the designer's task harder by placing restrictions on the available options. In reality, however, not all constraints originate strictly in the requirements specification . . . designers frequently impose constraints that are neither necessary nor objectively valid" (Gedenryd, 1998).

Clients provide constraints, but the most useful and actionable ones come from within the designer. For example, consider a designer who is developing a Web property that will offer streaming video content. The client has specified some requirements: the site needs to offer different content to paid members and visitors, and the content needs to be presented in HD quality. These requirements act as preliminary constraints. The designer begins to research video consumption on the Web, first through secondary research and competitive analysis and next by observing people using online media sites. This research produces a great deal of data, which contains additional constraints. Yet no individual user said, "Your design must allow me to easily embed this video on my Facebook profile" or "Your design must allow me to set parental controls." Instead, within all of the aforementioned research, these constraints lie hidden.

A designer may have watched someone viewing a clip, and the viewer may have said, "My friend Jimmy would think this is hilarious"; the designer may have watched another user alt-tab between a video site and Facebook; or the designer may have shared content with his own friends. All of these experiences point toward an implicit connection—video sharing through direct embedding—and this becomes a constraint or an opportunity for design.

The designer may have observed users at a coffee shop entertaining their children with a netbook. The designer may have just read an article about TV ratings. Or the designer may have personal experience of coming across undesirable "mature" content. All of these experiences also point toward an implicit connection—parental controls with default settings—and this, too, becomes a constraint or an opportunity for design.

In both cases, the constraint is implicit in the data, yet once it becomes explicit through action, the designer can embrace it as objective criteria for success. When the client asked the designer to create something, the something was vague and amorphous. These new elements add structure and tangibility to the creation and become points of departure for the creation of the site.

Constraints are flexible over the length of a project. Like other things that flex, however, they will snap if pushed too hard. In addition to designer-imposed constraints, clients also offer prescriptive boundaries around design projects. Client constraints can be selectively ignored only when the resulting design is successful, and that's a Catch-22 for designers. Obviously, designers cannot guarantee success, even when the work is informed by intuition, yet designers also cannot know whether they will succeed until they try. An experienced designer develops an ability to play the politician—to explain away why constraints were bent or broken—in an effort to bring a design idea to fruition and allow it to be tested.

Constraints present an interesting client challenge; they act as guidelines, not rules. Consider the following dialogue between a designer and a client concerning a Web-based flow:

> Designer: "This is such a critical moment in the checkout flow that I highlighted the area in red and made the action abilities a bit larger than on other pages."

> Client: "But that doesn't fit within the parameters of the templates we've established and everyone has signed off on. It's different from the other pages. Won't it be inconsistent?"

> Designer: "Yes, it is inconsistent. But I think, at this part of the flow, it's important to call it out as separate."

Client: "I'm confused. When we developed the templates, you told me consistency was important. Now you are telling me it's not important. Which is it?"

This is a difficult question for the designer to answer because the client is entirely right. Consistency is important, but as it is a constraint, it can be selectively ignored. However, the designer needs to understand how to describe this inconsistency, often by showing the benefits and user value of such a level of subjectivity. Unfortunately, this takes a substantial amount of effort, and it is often easier to simply revert to more accepted, "scientific" approaches: consistency, linearity, and a rigid embracement of constraints and requirements.

Being Playful

The ability to "be playful" is critical to achieve deep and meaningful synthesis of disparate ideas. Playfulness can be thought of as a casual, almost apathetic, work perspective, but the apathy is to the immediate significance of a particular design move, not to the long-term consequences of a particular action. That is, one might playfully suggest a design decision that is out of the question, and an objective viewpoint would consider this a form of destructive or deviant behavior. Yet this ability to suspend caring—to be apathetic in the moment but care passionately about the outcome—is critical to finding meaning in the chaos of design research.

To understand what is meant by "play" in design, it is useful to describe what is *not* meant by this word. When considering a playful environment, many immediately think of Nerf darts flying over a design studio, with twenty-somethings playing video games and sitting in beanbags. This unfortunate representation of the dot-com boom-and-bust era has tainted the notion of a creative environment by implying that being creatively playful requires a sense of slapstick humor and a lack of focus.

In fact, being playful is much more of a cultural *mindset* than a cultural *physicality*. Playful approaches to design can be found in the most boring of physical environments, simply because the administrative powers have allowed and encouraged people to explore, break the rules, and reject the "way things are normally done." Thus, playfulness in design has little to do with toys, props, or even physical environment and everything to do with the established political and cultural constraints.

To hold a playfully deviant point of view in the context of a serious design discussion allows a designer to explore divergent ideas, temporarily move the problem constraints, and expand the boundaries of what might be considered "appropriate" design decisions. Consider this example, taken from a real design consultancy that was brainstorming features for a real mobile-phone application:

Designer 1: I think our solution has to take advantage of GPS because location is so critical on a mobile. It's like . . .

Designer 2: Yeah, since it knows where you are all the time, it could let everyone know . . .

Designer 3: It could post your movement during the day and start to learn about you . . .

Designer 1: Yeah, finding a way to show where you go all the time . . .

Designer 2 [laughing]: Perfect to find the cheating boyfriend!

Designer 3 [also laughing]: You could set alerts, like "If you were in her house for more than 5 minutes . . ."

Designer 1: Nice, the "Catch Your Husband in the Act" mode. [Writes it on the whiteboard.]

Designer 2 [with a thoughtful look on her face]: What if we overlaid everyone's cheating to find the hussy of the neighborhood?

Designer 3: Nice, we could start to see location patterns, of which houses are getting the most traffic at weird hours of the day . . .

During this brief exchange, the designers explored what would obviously be a poor feature to include in a mass-produced product and an impossible feature to present this way to a client or executive sponsor. Yet the designers appear to seriously consider the feature, even giving it a name and writing it on the whiteboard. The designers have temporarily pushed the boundaries of what is "acceptable" in a product and have landed on an interesting new idea—the overlay of traffic to illustrate larger patterns of use. In fact, this idea—removed from the larger "Catch Your Husband in the Act" feature—finds its way into the actual product, allowing users to aggregate nightlife traffic to understand where interesting or fun events are drawing a crowd.

This exchange is typical of divergent thinking in a brainstorming context, and it relies on two main ideas. The first is that the designers feel comfortable enough with each other to propose outlandish, even offensive ideas that they logically know will never make it to market. They need to understand each other enough to embrace this type of play, and they need to see this type of thought as a competency, not as a disruption. To put a "crazy" or "bad" idea out in the world is to open yourself to criticism and ridicule, yet had Designer 2 not felt comfortable saying out loud, "Perfect to find the cheating boyfriend!" Designer 3 would not have had an insight about pattern finding. Imagine the awkward silence if the designers had not been at ease with each other.

The second enabling idea is that the company at which the designers work has established a culture that sees the value in this type of discussion, rather than writing it off as a "waste of time" or a "bad use of resources." During the earlier conversation, the designers were not answering e-mail, participating in a conference call, sitting at their desks, attending a formal meeting, or explicitly generating revenue. Not only that, but they were coming up with ideas that almost never would end up in the final product. When viewed through only a short-term lens of management, this is a "bad use of time." To understand the value of this type of playful thinking, one needs to take a longer view of product development. Unfortunately, this longer view is rarely afforded to most middle management in large corporations, and so a conversation like the aforementioned one is not likely to take place.

Alex Osborn, arguably the inventor of "brainstorming" as a technique, has been quoted as saying, "It is easier to tone down a wild idea than to think up a new one." As Osborn describes, brainstorming is a method of generating as many ideas as possible. The goal is not, as is popularly applied, to come up with "the best" idea. In fact, Osborn's formal method permits no judgment during the actual "storming"; only after numerous ideas have been developed does the group prune and qualify them. In this way, playful ideas like the one mentioned earler are encouraged, and the cultural attitude toward idea generation is supportive and receptive (Osborn, 1963).

Psychologist Joy Paul Guilford formalized Osborn's views by coining the phrase "divergent thinking" as a cognitive psychology phenomenon. Guilford recognized that intellect is multidimensional, and that some people are better at thinking abductively (and broadly), while others are better at thinking deductively (and narrowly). It seems that, at one point, nearly everyone was capable of abductive thinking as a default manner of looking at the world. Epistemologist Jean Piaget's constructivist theory implicitly accepts that a child actively learns through a form of divergent thinking and play, and Montessori education formally embraces this idea that physical play translates into intellectual progress. By contrast, rote methods of learning emphasize an outcome rather than a process. It might be only through these methods that most people have learned to focus on a singular "right" answer instead of thinking of multiple ideas.

The notion of being playful is to appreciate and encourage divergent, abductive thinking and to encourage the shifting, flexing, and removing of constraints and the exploring of "what-if" scenarios; that is, dream states. Our lives, jobs, and compensation are so frequently tied to rational thought that we have often forgotten how to actively dream, yet these dreams—the ability to generate ideas, outlandish or otherwise—are at the core of design innovation. Design synthesis embraces this divergent dreaming.

Author Edward de Bono taps into the divergent nature of play with his "six thinking hats" method of creativity. De Bono asks that designers embrace six ways of thinking by putting on (literally or metaphorically) six different colored hats, each representing a different way of approaching new ideas:

1. White hat—neutral and objective, concerned with facts and figures

2. Red hat—the emotional view

3. Black hat—careful and cautious, the "devil's advocate" hat

4. Yellow hat—sunny and positive

5. Green hat—associated with fertile growth, creativity, and new ideas

6. Blue hat—cool, the color of the sky, and above everything else, the organizing hat

As de Bono explains, the green hat is for lateral thinking, a phrase he developed to describe "pattern switching in an asymmetric patterning system." The green thinking hat allows people to break out of their understanding of what is and what should be and consider what might be—through humor and play. De Bono goes on to describe "a very close relationship between the mechanisms of humor and the mechanisms of lateral thinking. Both depend on the asymmetric nature of the patterns of perception. This is the basis of the sudden jump or insight after which something becomes obvious" (de Bono, 1999).

Our ability to be playful—even if only temporarily—is critical to developing new ideas, allowing them to live long enough to engage this pattern switching, and to cultivate even *more* ideas.

Experiencing Flow

Although psychologist Mihaly Csikszentmihalyi was not the first to observe that artists, designers, and other creative folks engage in a type of meditative state when they make things, he was instrumental in carving out a particular area of cognitive psychology called "flow." According to Csikszentmihalyi, flow is an optimal experience achieved during creativity that is an "automatic, effortless, yet highly focused state of consciousness" (Csikszentmihalyi, 1997). This optimal experience is one that can be identified in art, design, writing, cooking, sports, and even driving—any activity that demands a sense of craft, expertise, and ability to master. Csikszentmihalyi defines four specific attributes of the flow experience that are central to a discussion of synthesis. These attributes describe the ideal conditions for finding the interpretative "spark" of making meaning out of complicated data.

1. *There is immediate feedback to one's actions.* Synthesis, when externalized, creates a constant conversation where each "move" illustrates a new state. Compare this to attempting to synthesize large quantities of data in the head. While one might be able to combine, rearrange, and try new organizational techniques implicitly, the lack of immediate feedback makes it difficult to understand when progress has been made. Additionally, the feedback is quickly and implicitly critiqued, and corrected. Csikszentmihalyi describes that "in a flow experience we know how well we are doing. The musician hears right away whether the note played is the one." Similarly, in a synthesis exercise, a designer sees immediately whether a move is "the one."

2. *Action and awareness are merged.* Synthesis requires that a designer attend to the subject matter at hand *completely*. Practically, this requires a room with no distractions, a blatant rejection of e-mail, a turning off of mobile phones. The problems and realities of "real life" need to be temporarily ignored to permit immersion in the problem space. Then actions can be performed in the context of the gathered data, and not negatively impacted by people coming and going, or by meetings or conference calls. In addition, actions are made with an expert confidence. The designer is not held back by a lack of skill with a particular tool. The tool acts as an extension of the mind, limited only by the interim translation of the hands or eyes.

3. *There is no worry of failure.* As Csikszentmihalyi describes, "While in flow, we are too involved to be concerned with failure." To some degree, a designer needs to trust that synthesis will occur when appropriate methods are used because this trust in method and process will allow for a break from self-critique and from a meta-evaluation of the experience. Constant introspection related to failure leads to a hyperawareness of technique. For an inexperienced designer, this constant reassessment of the technique can completely halt progress, resulting in a lack of progress and a misuse of the most valuable resource for synthesis, time.

4. *Self-consciousness disappears.* "In flow, we are too involved in what we are doing to care about protecting the ego." Perhaps most important, this characteristic of design synthesis affords the ability to try new organizational schemas, to look at data in a new way, or to combine disparate data in "what-if" scenarios that may or may not succeed. Because synthesis is generative, even "wrong" synthesis approaches turn out to be "right" because they have allowed the designer to gain insight and knowledge about data and information.

During the spark of synthesis, these four attributes create an environment of meaning making. They create an atmosphere that embraces the previously described elements of abduction, constraints, and play. Flow is literally the awake-dreaming state of mind that occurs when a designer is able to move through the space of a problem, holding many design "moves" in the mind at once, and suspending self-criticism while retaining idea-based judgment.

As an example, consider a designer who is creating a strategy presentation for an executive review. This presentation will describe a vision for an entire product line, 18–24 months in the future. The designer has a deadline, knows the audience, and must work with a tremendous amount of content. These are the beginning constraints.

Consider three worst-case scenarios:

The designer is plagued with self-doubt. "What if I fail? What if I do it wrong? What if I can't figure out what to include in the presentation?" This self-doubt acts as a damper on divergence. The designer cannot suspend disbelief long enough to create something and react to it. Without something concrete, there is no interpretative meaning making or ideation. Although a designer may make things while troubled by self-doubt, the things themselves will likely be predictable, expected, and—most unfortunately—incomplete. Fear of failure will prohibit the act of completion, and no idea can then be used for internal reflection.

The designer internalizes the entire problem. Although the designer spends the entire day thinking hard about the problem, she makes no visual, tangible, and actual evidence of forward progress, so she receives no feedback of action. Unable to make concrete what is ethereal, the designer is constrained by the limits of the brain: a small short-term memory store, the inability to hold and examine multiple and competing ideas in depth, and the quick decay of what could be strong alternative new paths of ideation.

The designer is not good at the tools—or thinks he is not good at them. In synthesis, the tools are artificial constructs. Yet the methods of creating these artificial constructs are poorly communicated from one designer to another, and a blank canvas can be more daunting than the most complex computer program. A designer who thinks he is not good at creating the necessary models and constructs will become unable to merge action and awareness. A movement will be too conscious, or too forced, and the brain will focus more on the action and less on the problem at hand.

Experienced designers have embraced shortcuts to allevi-ate some of these problems. Some designers will force themselves to write their every idea, regardless of the imme-diate bearing on the problem at hand, on individual post-it notes. After 30 minutes, they have overcome the empty-can-vas problem and externalized whatever doubt they may have about their own abilities to solve the problem. In a sense, they force away whatever barriers to flow exist. This is a method of synthesis known as Affinity Diagramming, and it is described in more depth later in this book.

Other designers will schedule three or four hour-long work sessions in a closed conference room with other creatives, then ignore all "best practices" for successful meetings, such as agen-das, action items, and goals. Instead, the designers will work on a whiteboard to "talk out" the problem, without worrying about wasting time or approaching the issue from the "wrong way." This serves as a method of externalization, albeit through conversation rather than formal process.

Using Visualization as a Primary Mechanism of Thought

Unlike our long-term memory, in which we can store infinite amounts of material, the capacity of our short-term working memory is limited. This limitation, or "bottleneck" as Philip Johnson-Laird describes it (Johnson-Laird, 2009), impedes all forms of logical thought. Visualization is the way around the bottleneck.

Put simply, visualization is the act of externalizing ideas. Visualization can be formal and time consuming, as in the case of a high-fidelity computer rendering, or it can be quick and messy, as in the sketch on the back of a napkin. Both serve to sidestep the limitations of working memory, offering tre-mendous benefits for the process of synthesis. These benefits are described next.

1. Visualization allows for comparison. Generally, the unaided human brain can compare and contrast only small quantities of things and attributes. A visual representation of the comparison greatly aids

this process. Try it. Compare the following ideas, first without visualization and then with the diagrams in Figure 5.1:

Picture two fields of crops. The first field is divided into 16 sections, each of equal size. Five different crops are planted in the field: corn, peppers, lettuce, tomatoes, and beets. Each section is limited to one crop. There are twice as many peppers as beets, and twice as much corn as peppers. But in the second field, the crops are planted in equal amounts. Which field has more peppers?

The visualization not only makes the solution visually clear, but it also offers other new data (such as proximity) that are not present in the original presentation of the problem.

2. Visualization can easily convey how changes occur over time. A single visualization can be repeated, with the majority of elements staying the same and a few elements showing changes over time. Again, although this can be conveyed through words, the mind has a difficult time retaining the original state, thus making a comparison extremely difficult.

3. Visualization can be used to describe spatial relationships, helping to translate data into a perceptual form for different, and sometimes easier, recall.

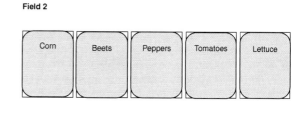

Figure 5.1
Which field has more peppers?

4. Visualization makes concrete what was before only an idea. This "realness" affects the way we consider an idea: It makes it sharable, storable, and critiqueable. All of these qualities contribute positively to design synthesis.

5. Perhaps most important, visualization allows our brain to build connections between disparate ideas, making meaning and creating new knowledge. This is related to the perception acting as an "afterimage" for working memory. Author Jonah Lehrer explains how this works on a physiological level. He explains that "... studies show that neurons in the prefrontal areas will fire in response to a stimulus—such as the sight of some cockpit instrumentation—and then keep on firing for several seconds after the stimulus has disappeared. This echo of activity allows the brain to make creative associations as seemingly unrelated sensations and ideas overlap" (Lehrer, 2009). The prefrontal areas of the brain allow us to take learned ideas and principles and apply them in new contexts. Visualization helps us to essentially offload the data from working memory, so other ideas can enter working memory and be "mingled" and synthesized.

Visualization is not reserved for those who "can draw," because a visualization need not be either representative or refined. A sketch on a napkin or a series of bubbles on a whiteboard are the only types of visualizations necessary to achieve the aforementioned benefits.

Changing a Prohibitive Culture

Given the importance of a supportive and encouraging culture, why don't more people and companies embrace these attributes during the development of new products? Simply, the traditions of most businesses do not support these qualities, and there has been no explicit effort to change these traditions.

Consider the manner in which most large enterprise product managers work. They spend most of their time in meetings, and a great number of these meetings are held over

Exposing the Magic of Design

the phone or via a remote conferencing system. There is often an agenda, and the meeting evolves in a fairly predictable fashion. After a number of people join and announce themselves, the meeting owner describes the objective of the hour. Various people add their comments and color to the discussion, and notes are taken. Several people will arrive late, disrupting the meeting and requiring backtracking. Often, someone will be asked a question, and after an uncomfortable pause, the person will say, "I'm sorry, can you repeat the question? I was multitasking." Someone may try to share something via WebEx or Live Meeting, and the technology will likely fail. Eventually, the meeting will end—not because the goals of the meeting have been met, but because the allotted hour is over. The meeting notes will be distributed via e-mail, and few will actually read them. The scripted manner in which these types of meetings unfold would be comical, except that these meetings are the norm and not the exception in corporate America.

Compare this scenario just described with the qualities of play and flow. These qualities demand spontaneity; humor; an amorphous, ill-defined sense of time; and a lack of interruptions. Being playful does not follow a timeline ("Be playful—now!" does not work), and one cannot ensure that flow will be accomplished during a 55-minute session. The structure of corporate meetings negates the ability to enact these qualities in a meaningful way.

Other qualities common in corporate product development are the ideas of "socialization" and "consensus building": that nothing can be considered done or can move forward in a substantial way without input and signoff from a larger group of stakeholders. Yet by its very definition, abduction demands that something is posited as correct *and built upon*, that groundwork is defined to see what happens when a particular direction is followed. Essentially, abductive thinking demands that a team try something, even if that something is a failure, to inform subsequent approaches. Socialization puts the breaks on an idea before it gets too far along, lest a stakeholder feel slighted that his opinion was not considered. An idea might have been developed with Edward de Bono's green hat on, but it is socialized by those wearing the logical and pessimistic black hat. The idea certainly will not get very far through this process. Because socialization and abduction are at odds, it is exceptionally difficult for large enterprises to develop any traction toward "innovative ideas" during the preliminary phases of product development.

Additionally, many large corporations have created a culture that rewards success and penalizes failure. Appropriately, those who succeed have learned to minimize their potential for failure and not to take great risks. This means doing things that are more conservative and putting in place large systems of planning, organization, and risk management. Often, these corporations attempt to think through every detail of a new product prior to making anything. A functional specification document commonly describes every feature and function in a product before it is built. Yet the functional specification rejects the process of generative and abductive thinking and ignores the role of *making* in the process of *thinking*. As Roger Martin, dean of the Rotman School of Management at the University of Toronto describes, "As integrative thinkers put their resolution through multiple prototypes and iterations, they use generative reasoning—whose raw material, remember, is what does not yet exist—to work back down from resolution to architecture to causality to salience. Large organizations may not recognize generative reasoning as a legitimate mode of inquiry, but they depend on it for lasting competitive advantage" (Martin, 2009).

Finally, consider the idea of a requirement definition document being "baselined"—a common practice in large corporate organizations. This specification document explains, in excruciating detail, what a product will be and how it will act. The baselining process then forces any changes to this document to jump through a rigorous set of hoops. Yet the entire notion of constraints and purpose demands a sense of flexibility, as further knowledge forces the reinvestigation of prior guidelines. The structures put in place to support the management of the giant document temper innovative design thinking, because they penalize late changes even when these changes are for the betterment of the product and the users.

Exposing the Magic of Design

Section Three
Methods and Applicability

The spark of synthesis occurs in the context of the DIKW system—moving from *d*ata, to *i*nformation, to *k*nowledge, and then to *w*isdom. The path has been routinely analyzed in fields of Information Technology and Knowledge Management, and it is mentioned by designer Nathan Shedroff in a brief article titled "An Overview of Understanding" (Shedroff, 2000). As Shedroff describes, data alone have little value. Although *data* usually implies numbers, it simply represents discrete units of content, without context and with no organizational mechanism. Creating information out of data may seem a simple task: Determine the units of data that are relevant and remove the rest. However, determining what is relevant requires deep interpretation, subjective pruning, and a framework in which to identify a hierarchy of importance.

Information can be thought of as meaningful data. Information is the organization of data in ways that illustrate meaning. In fact, this organization may alter the meaning itself. This has an important implication, because the meaning of seemingly objective data is altered by the appearance and structure of that data.

Knowledge results from the combination of elements of information to arrive at a principle, a theory, or a story. Although information may be sensory, knowledge seems to be more complicated and relies on personal experience and tacit knowledge. Storytelling, with its long history as a mechanism of knowledge transfer, can be considered a rapid immersion in experience.

Often thought of as enlightenment, wisdom can result from applying knowledge in a new and novel manner.

This path from data to wisdom is not linear, and it is not nearly as "clean" as just described. But this path will occur implicitly during design synthesis as you attempt to organize, manipulate, prune, and filter gathered data into a cohesive structure for meaning making. This requires a number of tools and techniques, many of which are subjective. The subjectivity of the design process indicates that *the ability to replicate findings*—a critical aspect of a scientific method or process—is not a relevant part of design. One quite capable designer may synthesize gathered data with a certain result, whereas another equally capable designer may synthesize the same data with a wildly different result.

The synthesis phase of the design process requires the designer's ability to cross several increasingly difficult chasms in the pursuit of understanding. This pursuit is both selfish and altruistic at once. You will seek to achieve a sense of knowledge acquisition to adequately begin the process of ideation. Additionally, to encourage a sense of timelessness in the development of designed artifacts, you will have to encourage the same sense of knowledge acquisition for your end user. If the user has learned, or felt passionately, or experienced a positive (or even negative) interaction with an artifact, you will have created behavioral resonance. The three chasms separating data, information, knowledge, and wisdom are illustrated in Figure 6.1.

When you are considering your particular design problem, and thinking about which methods to use, Table 6.1 will help you find a method or approach that is appropriate.

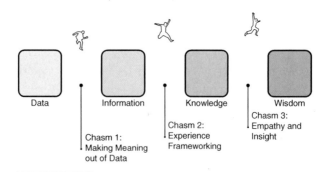

| Data | Information | Knowledge | Wisdom |

Chasm 1: Making Meaning out of Data

Chasm 2: Experience Frameworking

Chasm 3: Empathy and Insight

Figure 6.1
Chasms.

Table 6.1 A Method Selection Guide

Method Name	Cross the Chasm	Use This Method When:
Affinity diagramming	Make meaning out of data	You have a lot of data, and you want to create some form of organization. You are trying to identify initial themes or patterns. You need to abstract the complexity of your research to something more consumable, perhaps by an executive audience.
Flow diagramming	Make meaning out of data	You are trying to understand or visualize a sequence of actions. You are seeking to create an archetypical path through an interface or system. You are attempting to identify inefficiencies or opportunities in a process.
Concept mapping	Build an experience framework	You need to think through and illustrate a complicated idea in a simple way. You want to illustrate relationships between people, systems, or objects.
Forced semantic zoom (ecosystem mapping)	Build an experience framework	You are attempting to create an innovative product, system, or service. You want to understand and show the larger *context* of a single product. You want to describe brand relationships, rather than product relationships.
Forced temporal zoom (customer journey mapping)	Build an experience framework	You are attempting to create an innovative product, system, or service. You want to understand and show the larger *timeline* of a single product. You want to describe brand relationships, rather than product relationships. You want to identify opportunities to extend a brand into new territories.
Reframing	Create empathy and insight	You are attempting to create an innovative product, system, or service. You want to expand a product offering into new and unexpected areas.
Insight combination	Create empathy and insight	You are attempting to create an innovative product, system, or service. You want to develop a breadth of new ideas.

Chapter 6
Methods for Making Meaning out of Data

As you conclude the research phase of the design process, you will have generated a large quantity of data. The data, although potentially of great use, can be overwhelming and unclear. Add complex content and often scarce time, and you are forced into an immediate conundrum: You must make sense of the data by immersing yourself in it, but the immersion takes time—and the more time spent inspecting the data, the less time can be spent actually *designing*.

The goal of this immersion process, then, is to quickly make meaning and create information. Six immediate techniques will begin to help you understand how to do this.

Externalizing the Process (Get out of Your Laptop!)

The data created or gathered from contextual research will often take many forms: photographs, video clips, transcripts, requirements lists, magazine clippings, and other artifacts related to the problem. In an effort to maintain some sense of coherence, designers frequently horde the data in their laptops; the digital format lets them easily organize it into files, folders, and databases. But the file structure also arbitrarily limits the ability to manipulate individual pieces of data freely across file types, to form connections between pieces of data, and to manipulate the data quickly. The physical limitation of the laptop (the size), combined with the digital limitations of the software (the organizational schema), dramatically limits the designer's ability to understand the research in totality.

Synthesis requires that you forge connections between seemingly unrelated issues through a process of selective pruning and visual organization. Because of the vastness of data gathered while solving even a simple design problem, the quantity of data to be analyzed is often too large to hold in attentive memory at one time, so you must externalize the data through a process of spatialization. Among the few tools that designers continually use for this process are some incredibly effective ones: *a big wall, a marker,* and *a lot of post-it notes.* These tools will help you gain a strong mental model of the design space. The externalization of the research data allows for a progressive escape from the mess of gathered data.

Once the data have been externalized and the literal mess is begun to be reduced, you can begin the more intellectual task of identifying explicit and implicit relationships. Begin to physically move content around, placing related items next to each other. This process is less about finding "right" relationships and more about finding "good" relationships. All of the content is related in some way, but the important connections are frequently multifaceted, complex, and rooted in culture. Thus, it may be necessary to duplicate content (to allow it to connect to multiple groups) or to abandon or rearrange already established groupings several times during this process. This relationship-building task can occur over hours, days, or even weeks. Designers commonly explain that the discussion that occurs during this process is fundamentally more important than the output itself, because the conversation "fills in the gaps" in the research.

Once the groupings begin to emerge through the process of organization, labels can make both the literal and the implied contents of the group explicit.

Frequently, designers spend a great deal of time creating a war-room-style wall of data. This synthesis wall delineates the design space, allows for a collaborative process of sensemaking, and provides a spatial understanding of structure. Once the designers draw conclusions and progress through the ideation phase, the wall becomes unnecessary, so it is ignored for the rest of the project.

By taking the data out of the cognitive realm (the head) and removing it from the digital realm (the computer), and making it tangible in the physical realm (the wall) in one cohesive visual structure, you are freed of the natural memory

limitations of the brain and the artificial organizational limitations of technology. Content can now be freely moved and
manipulated, and the entire set of data can be seen at one
time. Implicit and hidden meanings are uncovered by relating
otherwise discrete chunks of data to one another and positioning these chunks in the context of human behavior.

Using Visual Design to Clean up the Mess

Once the content has been externalized, you are left with a
mess—quite literally. The sheer amount of data can take up
entire walls or offices, and the lack of consistency between
artifacts can look sloppy or disturbing. Some designers
describe a sense of anxiety from even looking at such a room.

Therefore, a second meaning-making technique is to use
principles of basic visual design and crasftsmanship—like horizontal and vertical alignment, attention to detail, and patience—
to begin to build uniformity into walls of artifacts. Use a
consistent layout, square off corners, and legibly rewrite key
points. Because content is always in flux during synthesis, you
might need several "cleanup sweeps" throughout the process.

Once the content groupings have been established, some
designers find it useful to plot large lists of words and data
with a digital layout tool and to put these large posters literally
on top of the gathered data. Simply by glancing at that area of
the room, the designer can simultaneously see the data on the
poster and can easily trigger the memories and thoughts that
were present when the data was originally gathered.

Organizing to Produce Semantic Relationships

Once you have cleaned up the *literal* mess, you can begin the
more intellectual task of cleaning up the *theoretical* mess. Use
an organizational process to identify the content elements'
explicit and implicit relationships, including patterns made by
combining the elements. Identifying a relationship forces the
introduction of a credible and abductively logical, although
often unvalidated, story of why the elements are related.

As new elements (gleaned from past experiences in your
life) combine with existing elements, the activity of defining

and forging connections actively produces knowledge. This process is best performed as a group, because the most important part is the conversation related to the story.

Prioritizing the Data to Emphasize What Is Important

As previously described, research produces a large amount of data. During the process of synthesis, you must decide that one piece of data is more important than another. This is accomplished by using an often implicit scale of importance by which to compare the data. The scale of importance is subjectively, but reasonably derived; its use is generally objective (within the system, each element is consistently compared). Data prioritization eventually identifies multiple elements that can be seen as complementary, thus creating a hierarchical data structure.

Judging the Data to Reduce the Quantity

But not all of the data identified in a discovery process is relevant. The process of synthesis forces you to consider relevance, because you pass the gathered data "through a large sieve" to determine what is most significant in the current context. Synthesis methods, then, require a constant reassessment of the current state as compared to the unknown end state.

Enhancing the Data through "Best Guess" Intuitive Leaps

Having organized these interpretations of the content, now you must begin to visualize and question them. You must begin to ask "why" and do your best to answer the question credibly and abductively. Often, these visualized answers take the form of charts, diagrams, graphs, or sketches. Because the output is only a guess, initial visualizations are better produced at low fidelity, with a large marker on paper, rather than investing a lot of time on pixel-perfect diagrams. The low fidelity of these sketches also will help you discard them when they prove to be the "wrong" solution, because you will not have invested enough effort to become emotionally attached to the idea.

Exposing the Magic of Design

For Example:
Getting to Meaning through Story
Contributed by Rachel Hinman, Nokia

We'll never know how the world "really" is. We'll always have to construct what we think the world is . . . and humans do this by telling stories.—Jerome Bruner

The goal of almost any designer worth her salt is to create work that has value to others: to design products and interfaces that will contribute to the human experience in a positive and constructive way. But how? How does a modern designer working in a diverse, globally connected world ensure that her work that is functional, emotionally resonate, and most important, meaningful?

Over the last decade, qualitative design research has been lauded as the "secret step" in the design process that guarantees this kind of success. Field research is supposed to yield the kind of information that brings design teams to meaning. But why?

Why does talking to people and listening to their stories of their broken relationship with their television or the guilt they feel about declining a friend request on Facebook make our design work better?

And how do design teams translate those hours of conversations into something actionable?

Different people have different answers to these questions. My answer is found in the stories and in a simple process of synthesis. I identify *what we saw and heard, what it means,* and *why it matters.*

The Why: Understanding through Narratives. One of my design heroes is not actually a designer; he is a cognitive and educational psychologist. In the early 1990s, Jerome Bruner published a seminal article entitled "The Narrative Construction of Reality," in which he argued that people structure their sense of reality through narratives. In the article, Bruner maintains that people make sense of the world by identifying similarities and differences. He argues that the human mind structures its sense of reality by reconciling the similarities and differences in cultural products such as language and symbolic systems. Of particular interest is that people describe these similarities and differences through *narratives*.

To get a better sense of this theory in action, try asking friends or colleagues about their experience using the remote control of their television set. Here is my mom's response:

> *I actually have two remote controls—one for the TV and one for the satellite dish. Using them is like opening a padlock. There's a special sequence you have to push the buttons in and if you don't do it in the right order, it doesn't work. Then if you're not close enough to the TV, it doesn't work either. So I usually have to dance around the TV like a goof ball for a couple minutes with two remote controls just to watch Law and Order. I don't know why there has to be so many buttons and why they have to make it so complicated—I am turning on a television, not landing a plane.*

Note how my mom describes the experience by comparing it to several other activities that people can relate to (opening a padlock, dancing around, landing a plane). In this simple quote, there are a lot of inferences on how to improve the design of remote controls.

Bruner maintains that humans learn to do this activity of comparison at a very young age. In his book *Making Meaning*, Bruner references a simple and elegant experiment in which kindergarten children were asked to look at two photos. Both photos depict a young girl at her own birthday party. In one photo, the birthday girl is smiling and happy. In the other photo, the birthday girl is crying and is clearly displeased. The children were then asked to select a photo and explain what is happening in the photo. Interestingly, each child in the study chose the photo of the girl crying. One child explains the tears were caused by poor party attendance; nobody showed up. Another child postulates the tears were caused by the girl arguing with her mother over which party dress to wear. Another child claimed the tears were caused by not receiving the birthday present she had hoped for.

Bruner's experiment underscored the fundamental difference between definition and meaning. The definition of a birthday party is an event to celebrate one's birth. But what a party means—the cultural expectations, the social script, the attributes that define the experience—is described through story. People use narratives to describe expectations. The children all selected the photo of the girl crying because it defied what they would expect to see at a birthday party. Their stories of poor party attendance, mothers enforcing bad clothing choices, and crappy presents mirrored the process by which our minds reconcile similarities and differences in expectation. The children were using narrative and comparison to describe their expectations of a birthday experience.

The How: Applicability to Designers. Design is about creating a product, services, or experiences that map to user expectations. Designers often rely on instincts to intuit these expectations. However, as the dimensions of user groups—dimensions such as culture, gender, and age—become more granular and distinctive, it becomes more important for designers to draw from real data to match the expectations of the people they design for. There is no better way to understand these expectations than to get out in the field and listen to the stories and perceptions of real people.

Although qualitative research has multiple benefits, it is time and labor intensive. Most notably, the analysis and synthesis of qualitative data can be an arduous task. The completion of field interviews leaves most design teams with a colossal mountain of videotapes, transcripts, and digital photos. The task of translating it into thought-provoking and actionable insights can cause even the most committed user-centered designer to run for the hills. I have done my share of floating in an abyss of qualitative data with seemingly no shore in sight. To save my own sanity, I developed the following framework to guide myself through analysis and synthesis.

What We Saw and Heard: Memoing. Part of what makes data analysis and synthesis tough to tackle is the fact that there is a lot of data, and it is in disparate forms. Slogging through it all can feel intimidating because there is simply so much *stuff*. This phase—the analysis and synthesis phase of the design process—has always seemed eerily like *Clean Sweep*, a television show about homeowners who have let their pack-rat tendencies go awry. Their homes are so full of clutter they are uninhabitable. Homeowners on the show must first unload the entire contents of their home into their front lawn. This is effective for two reasons. First, it forces people to face the madness and take a good hard look at all of their belongings. Then, it presents the clutter in an apples-to-apples format, where items can be compared and contrasted.

Similarly, it is important to examine field research's often-disparate data forms (video tapes, transcripts, digital photos, etc.) in an apples-to-apples format by transforming it into something tangible, visible, and consistent. Only then can you take a good hard look at everything that has been collected and start making hard choices about what is important.

Making Your Data Visible. Making sense of data starts during the debriefing of every interview. Different people have different perspectives on "note taking" during an interview. I do not take written notes during the interview because I find it distracting for the participant and for me. Instead, I videotape and audiotape the interview, and I schedule a half hour after an interview to debrief and capture initial impressions with the interview team. We usually go to a coffee shop and talk about what stood out for us from the interview. Soon we have a stack of post-it notes filled with quotes and keywords. Each post-it has a thought written on it, something like "called customer service every day for a week" or "pissed off when blood sugar is high or low—what am I doing wrong?"

At some point after all of the interviewing, I also review the audio or videotapes and see whether I missed anything. Whenever a participant says something I think is interesting or important, I write it on a post-it note and record the time-stamp from the video, which makes editing much easier. This process of recording things on post-its is a hybrid of Grounded Theory, a data-analysis process used by some anthropologists. The act of taking down important points from participants' conversation is known as "memoing" (see Fig. 6.2).

The "Why" of Post-It Notes. Some researchers like to work digitally, entering all their memo information in tools like Excel or Word. I find this challenging. The goal for ethnographic field data analysis is to build a shared understanding with a team. That requires collaboration. Computers make it difficult to share data and facilitate conversations around it. Unlike data trapped in an Excel spreadsheet, the tactile quality of post-it notes is easy for people to scan and engage. Memoing on post-it notes also provides a shared sense of ownership in the process, making it easy to build upon the ideas of others. Of course, post-it notes also have downsides: You cannot e-mail them or easily share them with teams working remotely. But I think the benefits of easy collaboration far outweigh the trade-offs.

Figure 6.2
An example of memoing.

Participant Boards. It is very easy to generate more notes than you know what to do with. I use the participants themselves as organizing principles. I dedicate a large sheet of cardboard, a wall, or window to each participant in the study. The team fills each space with the post-it notes and photos from the interviews. Each space functions as an initial organizing principle for the data, a small homage to each participant that highlights the important quotes and stories they have shared (see Fig. 6.3).

What It Means: Getting to Insights. Once all the data have been memoed and organized, the next step is to cull through the stories and data to find insights. I start by looking at the participant boards to identify commonalities in the memos that simply stand out as interesting or important. Collaborating with my team and my clients, I begin to group the memos into affinity clusters. The quotes and stories themselves become fodder for conversations, diagrams, and frameworks.

Collaboration is critical during this phase of the process. Sitting in a room by yourself with all those notes will drive you mad and make your results one-dimensional. Encourage your teammates to marinate in the data with you. Compare stories and scan for patterns. Discuss different points of view. Encourage debate. Most important, capture everything. Do not let your team conversations disappear into the ether.

Figure 6.3
An example
of a participant's space.

Exposing the Magic of Design

This "what it means" phase is about identifying the important insights in the stories you have heard. Just like the clutter bugs in Clean Sweep, you cannot keep all the data—there is too much of it. Through conversation and collaboration, you will find the gem-like insights that are most salient for your project.

Why It Matters: Emerging Insights and Implications. The final step of the process is editing your insights into something that has relevance and expresses a point of view. The goal is to come up with emerging insights that support a design implication. Together, the emerging insights and design implications recommend the actions you think should be taken next.

Editing insights is challenging. Typically, teams have more than they know what to do with, and insights are not equally relevant to the project. So you must continually ask yourself and your team, "Why does this matter?" Another helpful technique is to share the ideas, themes, and affinity clusters with people who know nothing about the project. Sharing your thinking can help determine which ideas resonate. Fresh eyes also help in honing the "talking points" and crafting the story or its emerging insights.

Emerging Insights. Creating a strong and persuasive emerging insight is a subtle dance between exposing data found in the field (the "what we saw/heard" information) and your interpretation of that data (the "what it means" information). It requires ruthless editing to move from the whole to the five to seven most important data findings, which become emerging insights. Use the stories from the participants to support these insights.

Implications. Each emerging insight should have an implication—a clear directive to the client. Implications tell people why the research matters and how it is relevant to them. Effective implications reframe the problem in an interesting and thought-provoking way. They are sticky and memorable. And just like good design, good research implications also express a point of view, an essential part of ensuring actionable research. You have gone into the field and talked to real people who have shared a part of their lives with you. You have a responsibility to share their perspective, even if it is unpopular. Your research implications should tell their story and give your clients or colleagues clear and actionable next steps.

What research implications should not do is tell clients what they already know (such as "people have powerful relationships with their mobile phones").

Putting It Together. The process I have described has three general steps:

1. *What we saw and heard.* Put your data in apples-to-apples format through memoing. Create participant boards—an homage to each user—as a first step in organizing the data.

2. *What it means.* Marinate in the data and collaborate with your colleagues. Encourage discussion and debate. Identify patterns. Capture the initial insights. Write everything down!

3. *Why it matters.* Organize the memo clusters and initial insights into emerging insights. Edit by frequently asking, *Why does this theme matter?* Find your point of view and create implications, the clear steps the client should take as a result of your findings.

The sheer volume of data produced during qualitative field studies can leave you wondering whether it is worth it. It is. Products, services, and experiences designed using this type of process have an inherent sensitivity to user needs and expectations. Listening to the stories of users will help you get beyond obvious solutions. It is the work that will get you to the meaning.

About the Author. Rachel Hinman is a designer/researcher/strategist and a recognized thought leader in the mobile user-experience field. With more than a decade of design industry experience, she is a strong believer in approaching mobile design and strategy from an empathic, human-centered perspective.

Rachel's passion for people, design, and the belief that people can use technology to improve their lives has been the driving force of her career. Her passions and interests lie in the convergence space: creating experiences beyond the desktop. Rachel's innate sensitivity to people and culture have proven to be powerful skills, enabling her to lead successful research studies on mobile phone usage in the United States, Europe, Asia, and Africa.

Rachel writes and speaks frequently on the topic of mobile research and design. She is the creative force behind the 90 Mobiles in 90 Days Project, and her perspectives on mobile user experience have been featured in *Interactions Magazine, BusinessWeek, Wired, Wireless Informatics Magazine,* and the *Adaptive Path* blog.

Rachel received a master's degree in design planning from the Institute of Design in Chicago. Her clients and previous employers have included IDEO, Microsoft, Yahoo!, Mobile, and Kaiser Permanente.

Method: Affinity Diagramming

An affinity is a sense of closeness or likeness. When ideas are said to have an affinity to one another, it means that they share enough essence to be deemed related. An affinity relationship is subjective and requires interpretation. When considered through a broad lens, everything has a connection to everything else, so you must use a smaller filter when you try to identify the affinity of ideas.

An affinity diagram is a method of discovering relationships among ideas, and then emphasizing and highlighting their likenesses. The diagramming process itself is generative, in that it creates further knowledge about the ideas. The method's output then helps to visualize the larger forged idea patterns and connections. Finding likeness helps to organize and to manage a seemingly overwhelming task. From thousands of discrete data points, a theme or set of themes can be found that point to a more manageable set of patterns. An affinity diagram is a tool for the eye as much as the mind. Although the act of organization is a cognitively demanding task, the results are perceptual because the clusters indicate relative scale and provide visual structure.

An affinity diagram assumes that your raw data have already been gathered and, to some degree, formalized. However, it can be used with other generative design methods, such as brainstorming, to simultaneously develop synthesize ideas. But this method need not be limited to ideas—it can be used to organize and find patterns in any sort of data. Commonly, statements from user interviews are used as individual data points. Each line from a transcribed interview acts as a unique data point. In this way, 10 or 15 individual interviews can be mingled to find themes, and the resulting diagram helps the design team to rank themes based on prominence and size.

How to Apply This Method

1. Write each piece of data on a note card or post-it note. A "piece of data" can be a word, a phrase, a sentence, a picture, a movie, or any other discrete element. The small size of the card or note adds portability and

impermanence, which is critical for the fluid manipulation of ideas. Code each piece of data uniquely with a reference to the source material so you can trace each note to its source. For example, if the diagram is going to map interview statements, the interview may be labeled by number (i.e., "interview 2") and the line number labeled by number ("22"), so the unique note card would be labeled "interview 2.22."

2. Once the cards have been individually generated, spread them out randomly on a surface. Random placement ensures that the order of initial generation is not creating an implicit pattern.

3. Now, begin to physically move the cards around, looking for patterns and groupings. Each movement can be logical or emotional, but the intention should be made concrete through a verbal statement like "I'm moving *this* card next to *this* one because both describe a way of paying for items" or "I feel like both of these cards are discussing purchasing, so I'm going to put them together." Slowly and collaboratively, work through the cards, finding patterns and building connections. It may be tempting to attach a label to a group of cards, but wait as long as possible. Once a group gets a label, it becomes "concrete," so participants will be reluctant to disrupt the grouping in favor of a better combination.

4. As you work through the entire set of cards, patterns and groups can become large. Consider breaking a pattern of eight or more notes into separate, smaller, and more defined groups.

5. Once all of the notes have been placed in groups, conduct a focusing exercise. Label each group on a larger, alternate-colored card. This label, which should capture the group's essence, might be "Purchasing," "Exploring Checkout," or "Ways of Shopping."

6. Next, develop and articulate a two- or three-sentence description of the category. This description should

capture the theme of the grouping, allowing a reader to understand the thematic contents of the group without actually reading each note. As an example, the description for "Ways of Shopping" might be "This grouping describes a number of ways in which a consumer would shop for items. It includes some new ways of thinking about purchasing and checkout."

7. Finally, put the entire effort into a document in an outline format, including the name, the description, and the individual content elements. The combination of the visual groupings and the more organizational outline become the deliverables used to communicate what you have done and what you have learned.

While developing the affinity diagram, the designer has begun to move from data to information by adding contextual containers for discrete ideas. The context ties to both the actual contents of the notes and the larger worldview of the designer.

Variations on this method include grouping entirely in silence to avoid cross-designer idea "contamination," and doing the grouping alone before combining result sets.

For Example:
Parallel Clustering

Contributed by Colleen Murray, Jump Associates

What follows is an approach for synthesizing numerous insights into clear, prioritized directions.

Overview. Imagine you are working on a team-based research project. You have already amassed a large set of data and analyzed it. You now have a number of interesting insights and possible directions. Unfortunately, everyone on the team has a different idea of what is most important. And your team needs to converge on some winning concepts sooner rather than later. How do you synthesize multiple perspectives into a set of clear, prioritized directions? More important, how do you prescribe, rather than simply describe where to go next? For some, parallel clustering is the answer.

A Parallel Clustering Case Study. In 2005, a large, international client asked us to help them explore their emerging markets strategy. The project was focused on the BRICS— the developing economies of Brazil, Russia, India, China, and South Africa. The goal of our work was to understand broad cultural themes in order to identify business opportunities in these countries. Our team was made up of four hybrid thinkers who closely collaborated with our client to better understand the day-to-day needs of small and midsized businesses, and the cultural drivers that differ from those in the United States. The project data came primarily from secondary research and expert interviews.

Selecting Methodologies: Using Parallel Clustering to Quickly Prioritize Themes. We needed to find a set of actionable themes based on our research, yet after weeks of analysis, we had too many insights, too many interesting directions, and a lack of clear focus on what was most important. Clustering activities, such as affinity diagramming, got us only so far. It was time to turn insights into action. We decided to organize a parallel clustering activity to help us extract and prioritize our top five themes.

Although our clients did not participate in the activity, we were able to share our process during discussions with them after the fact. Having documentation in place to show them how we got from point A to point B was helpful in providing transparency and getting buy-in on the final themes.

Understanding the Process

1. Take time to reflect. (Activity type: individual; time: 15 minutes)

   On the BRICS project, our first step was to give each team member a short time to reflect on the insights each found most important. The short time forced us to "think with our gut." Because the team had been living and breathing this content for 7 weeks, this exercise forced us to synthesize quickly what had been forming in the backs of our brains throughout the entire project. We each produced a list of our own most important insights.

2. Get out all the possibilities (Activity type: group; time: 60 minutes)

   Once we finished our lists, we came back together to develop a visual representation of all the possible theme areas—a mindmap. First, we assigned one of us to facilitate and record the conversation. Then each of us shared our lists while the recorder created the mindmap by capturing our words on a large sheet of paper. At this stage, these tips helped us consider a breadth of possibilities:

   - *Avoid introducing recorder bias.* In our examples, the recorder captured similar topics near one another without imparting personal bias into the map by making linkages explicit. The recorder stayed objective, asking questions and getting confirmation from participants that their ideas had been accurately represented.

   - *Capture variations and explore all possibilities.* If someone thought that her idea was like something else that was already on the paper, then the insight was recorded in close proximity. These slight variations were important to capture, not just for content variation but also because capturing them helped everyone on the team to feel heard.

- *Illustrate frequency of idea generation as a mechanism for describing popularity.* When, as in many cases, more than one of us had the same insight on our lists, we only needed to capture it once. However, we tried to indicate that the idea had been mentioned multiple times by adding a running tally in the corner of the idea.

We concluded the group session by making a photocopy of our map for each team member to work with (see Fig. 6.4).

3. Make your own clusters.

(*Activity type: individual; time: 15 minutes*)

Next, the group agreed to find approximately five clusters of related insights, and we developed a color code to identify each person's work to come. Then, we split up and individually began the process of clustering and prioritizing. Using

Figure 6.4
This is the initial mindmap created by the team after compiling everyone's top themes.

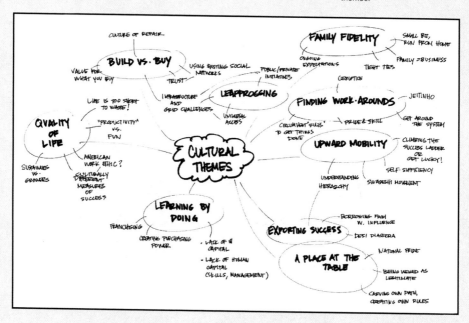

our signature color of marker, we each circled connections between insights that, as individuals, we saw as relating most closely, and we highlighted our top themes (see Figs. 6.5 and 6.6). The different colors later helped us to track who said what. Although the clustering step sounds simple, it occurs when folks are often most challenged. The step forces us to justify how we defined a particular space.

4. Compare outputs.

(*Activity type: group; time: 30 minutes*)

We came back together to compare our results. Taping all the mindmaps side by side on the wall, we compared what we saw. Clusters that were repeated across multiple maps were easy to prioritize as final themes. Most clusters had similarities with those of different teammates, but they were not exactly the same. We heard many comments such as "oh, I never thought of it that way." These differences revealed interesting nuances and valuable points of view. As we discussed further, we took notes to capture the insights, meanings, and decisions we made along the way.

5. Define meaningful, final themes.

Figure 6.5
We made a copy of the original mindmap for each team member.

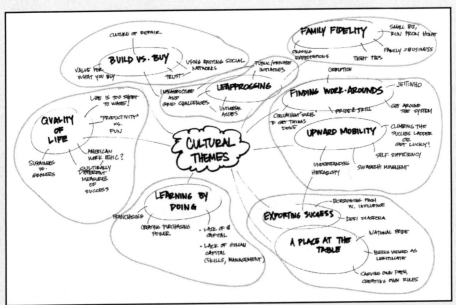

Exposing the Magic of Design

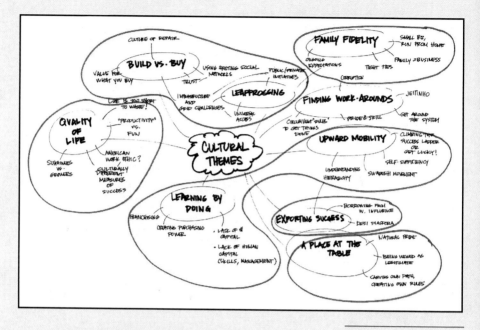

The mindmap contains the following handwritten labels:

CULTURE OF REPAIR

FAMILY FIDELITY — SMALL BIZ, RUN FROM HOME — ONGOING EXPECTATIONS — FAMILY = BUSINESS — TIGHT TIES — CORRUPTION

BUILD VS. BUY — USING EXISTING SOCIAL NETWORKS — PUBLIC/PRIVATE INITIATIVES — VALUE FOR WHAT YOU BUY — TRUST — INFRASTRUCTURE AND GRID CHALLENGES

LEAPFROGGING — UNIVERSAL ACCESS

FINDING WORK-AROUNDS — JEITINHO — CIRCUMVENT RULES TO GET THINGS DONE — PRIDE & SKILL — GET AROUND THE SYSTEM

LIFE IS TOO SHORT TO WASTE!

QUALITY OF LIFE — "PRODUCTIVITY" VS. FUN — AMERICAN WORK ETHIC? — SUSTAINED VS. GROWTH — CULTURALLY DIFFERENT MEASURES OF SUCCESS

CULTURAL THEMES

UPWARD MOBILITY — CLIMBING THE SUCCESS LADDER OR GET LUCKY! — SELF-SUFFICIENCY — UNDERSTANDING HIERARCHY — SWADESHI MOVEMENT

LEARNING BY DOING — FRANCHISING — CREATING PURCHASING POWER — LACK OF $ CAPITAL — LACK OF HUMAN CAPITAL (SKILLS, MANAGEMENT)

EXPORTING SUCCESS — BORROWING FROM W. INFLUENCE — DESI DIASPORA

A PLACE AT THE TABLE — NATIONAL PRIDE — BEING VIEWED AS LEGITIMATE — CARVING OWN PATH, CREATING OWN RULES

Figure 6.6.
These clusters demonstrate another person's alternative take on the most important clusters. Note the differences from Figure 6.5.

(Activity type: group; time: 30 minutes)

Next, we made final selections of themes. It was fairly painless because we had already spent the afternoon thinking about these themes, both individually and as a group. We had a good idea where everyone in the room stood. We knew what was important and what was not, and we knew different ways to cut the data. We combined all of our thoughts into a final mindmap that highlighted our final themes (see Fig. 6.7).

The Benefits of Parallel Clustering

Manage Complexity. Information overload can be a challenge for any synthesis activity. Parallel clustering forces each team member to quickly articulate the top ideas, themes, and needs they view as most important. This activity helps to cut through the clutter and very quickly identify what is most important.

Foster Collaboration. Most teams are made up of individuals with various personalities and working styles. Some folks are quiet. Others can dominate a group. And some people simply need more time than others to process what they think. Making everyone's thought processes explicit within a group can be a challenge. Parallel clustering gives everyone the time and space they need to share what they think is most

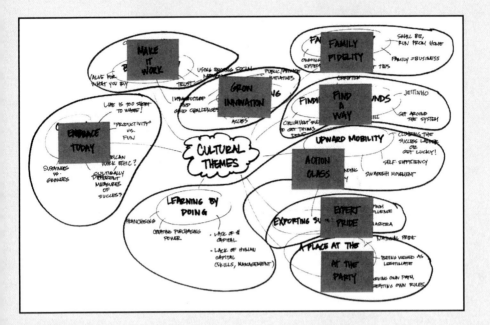

The following text appears within the figure image:

MAKE IT WORK · VALUE FOR WHAT YOU BUY · USING SPORTING SOCIAL MEDIA · PUBLIC/PRIVATE INITIATIVES · TRUST · GROW INNOVATION · INFRASTRUCTURE AND GRID CHALLENGES · ACCESS · LIFE IS TOO SHORT TO WASTE! · "PRODUCTIVITY" VS. FUN · EMBRACE TODAY · CAN WORK ETHIC? · CULTURALLY DIFFERENT MEASURES OF SUCCESS? · SUSTAINERS VS. GROWERS · CULTURAL THEMES · FAMILY FIDELITY · SMALL BIZ, RUN FROM HOME · FAMILY = BUSINESS · ONGOING EXTENDED · TIES · CORRUPTION · FIND A WAY · JEITINHO · GET AROUND THE SYSTEM · CIRCUMVENT RULES TO GET THINGS DONE · UPWARD MOBILITY · CLIMBING THE SUCCESS LADDER OR GET LUCKY! · SELF-SUFFICIENCY · ACTION CLASS · SWAPESH MOVEMENT · LEARNING BY DOING · FRANCHISING · CREATING PURCHASING POWER · LACK OF CAPITAL · LACK OF HUMAN CAPITAL (SKILLS, MANAGEMENT) · EXPORTING SUCCESS · EXPERT PRIDE · NATIONAL PRIDE · A PLACE AT THE PARTY · AT THE PARTY · BEING VIEWED AS LEGITIMATE · HAVING OWN PATH, CREATING OWN RULES

Figure 6.7
This is the team's final output of top themes. After a discussion, we agreed that this combination of eight clusters would best help us tell our story.

important and why. This helps teams to tap into unique individual perspectives and truly harness the team's wide array of experiences and skills.

See New Possibilities. Even when teams take the time to listen to one another, it does not mean that everyone hears the same thing. It can be difficult to understand others' nuances. But because these various flavors can add richness and depth to insights, systems must be used to capture the details. Creating a shared recording or mindmap helps to reveal both the similarities and differences in people's framing of topics. Physically clustering similar or overlapping ideas near each another quickly reveals what people agree on and what they do not.

Make Trade-Off Decisions. At its essence, synthesis is about combining two or more entities to form something new. There are many ways to gain clarity on decisions about which direction to pursue. A visual representation of the different ways that folks prioritize identical data points helps to foster a rich discussion on the must-haves, the nice-to-haves, and the different paths to those decisions.

Exposing the Magic of Design

Move Quickly. Get the team together in a room. Force people to "think with their gut." Prioritize what is most important. Get all the ideas out. Share as a group. Put these together, and you will move from chaos to clarity with great speed. With a little planning, the entire parallel clustering activity takes only 2-1/2 hours.

In Summary. The synthesis process for teams looks and feels like a unique mix of familiar activities; clustering, mindmapping, and idea ranking. Yet parallel clustering is different. Unlike a simple clustering exercise, parallel clustering quickly reveals multiple alternatives for grouping information. It is different from a final ranking or selection activity, because it builds robust themes, rather than simply filtering ideas out. And the technique is distinct from personal mindmapping, because shared maps make everyone's thinking transparent to the entire group.

About the Author. Colleen Murray brings an interaction designer's sensibility to the creation of compelling future strategies. Her design experience has helped her to better parse the relationship between people and the objects that surround them. A significant portion of Colleen's work has focused on helping clients to map out potential new business opportunities. She has helped Jump to develop proprietary methods for analysis and identification of promising growth areas. Her work has explored a number of industries, including consumer electronics, digital entertainment, and office environments. Colleen holds a master's degree in design planning from the Illinois Institute of Technology and a bachelor of fine arts degree in graphic design from the University of Illinois.

Method: Flow Diagramming

Ethnographers Hugh Beyer and Karen Holtzblatt developed a set of visual work models in their text *Contextual Design: A Customer-Centered Approach to Systems Design* (Holtzblatt & Beyer, 1997). The common visual style of the models creates a language for seeing and understanding work as it occurs, and the models can then be referenced throughout design. Although the intention of these work models was to help usability engineers create usable software, the visual style of making connections is ideal for designers focused on interpreting complicated data. More important, the act of modeling is a method of design synthesis. By producing work models, designers generate new knowledge, identify gaps in their shared understanding of a problem space, find new ways of looking at a problem, and help identify pragmatic and easily solvable design problems.

Beyer and Holtzblatt developed five models to capture the nuances of ethnographic data: the Flow, Cultural, Sequence, Physical, and Artifact models. The designed object can be introduced into the models and the effects this object would have on the entire work system can be predicted.

The Cultural model attempts to illustrate only interpersonal relationships and to articulate the effects of political or social factors. In this way, the model describes the human, emotional, and often subjective business environment that shapes the way design decisions are made. As an example, in many large companies, a single decision maker might feel a need to gain consensus among peers before making a design decision. This is a cultural feeling within the particular workplace, and it would be captured in the diagram (see Fig. 6.8).

The Sequence model shows the order of triggers, intentions, and actions that occur during work, allowing for an understanding of implicit task structure and hierarchy. This model breaks down work into a concrete and specific list of activities, describing the causes and results of each activity in a process (see Fig. 6.9).

The Physical and Artifact models (see Figs. 6.10 and 6.11) capture the built environment and the objects used to complete tasks, showing both a visualization of these items and the details of their use and misuse.

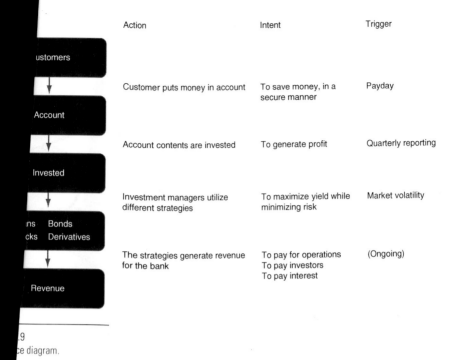

Action	Intent	Trigger
Customer puts money in account	To save money, in a secure manner	Payday
Account contents are invested	To generate profit	Quarterly reporting
Investment managers utilize different strategies	To maximize yield while minimizing risk	Market volatility
The strategies generate revenue for the bank	To pay for operations To pay investors To pay interest	(Ongoing)

ustomers

Account

Invested

ns   Bonds
cks   Derivatives

Revenue

.9
ce diagram.

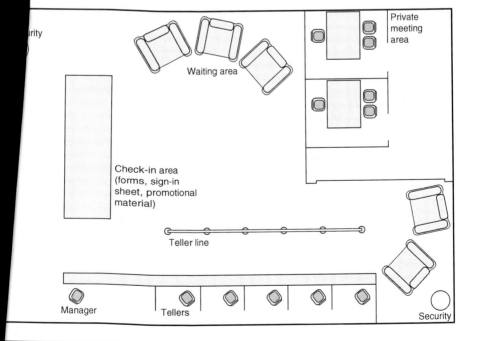

Figure 6.10
Physical diagram.

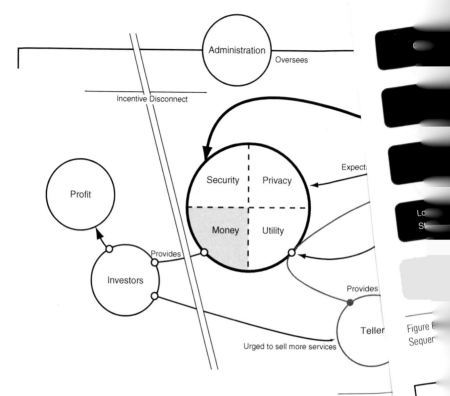

Administration

Oversees

Incentive Disconnect

Expect

Security | Privacy

Profit

Money | Utility

Lo
St

Provides

Investors

Provides

Teller   Figure 6
Sequer

Urged to sell more services

Figure 6.8   Sec
Cultural diagra

The Flow model (see Fig. 6.12) captures the movement of information without regard for time; it allows for a visual synthesis of data and information transfer. This model connects entities and actions, but because time is abstracted, the model acts as a conglomerate of activity, communication, and knowledge share. It is as though someone recorded each person individually in an organization to see what everyone was doing, then laid each frame from each recording on top of one another on a single diagram. The diagram shows handoffs of artifacts, describes where responsibilities begin and end, and illustrates the mechanisms and vehicles that are used to accomplish work.

Although all of the models are useful, the flow model is particularly relevant in capturing information related to synthesis and moving from data to information. The flow model captures the chaos inherent in large, intertwined systems. Although the data are interesting alone, the model provides a valuable context for this data by showing each piece as feeding into, or being fed by, a larger flow and transport of content.

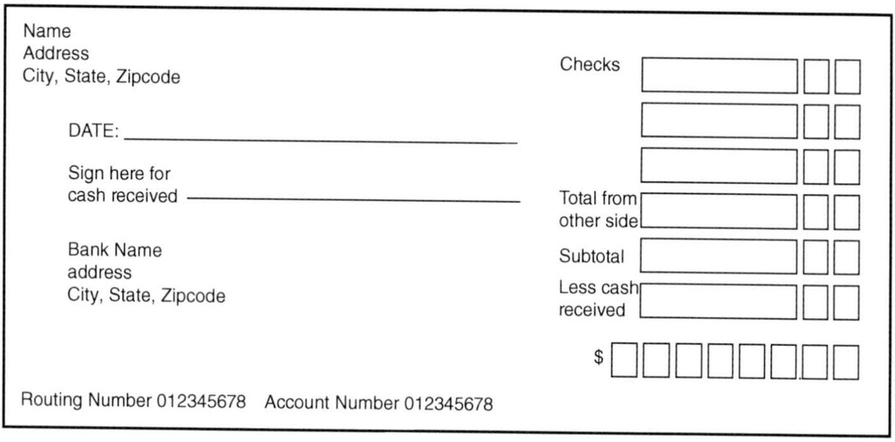

```
Name
Address
City, State, Zipcode                          Checks    [          ] [][]

    DATE: _____            [          ] [][]

    Sign here for                                        [          ] [][]
    cash received  _____   Total from[          ] [][]
                                                  other side
    Bank Name                                     Subtotal  [          ] [][]
    address
    City, State, Zipcode                          Less cash [          ] [][]
                                                  received

                                               $ [][][][][][] [][]

    Routing Number 012345678   Account Number 012345678
```

Figure 6.11
Artifact diagram.

The process diagram is a type of Flow model that illustrates the logical path through a system, showing decision points and articulating cause-and-effect relationships and system use. This diagram (also known as data flow diagrams or decision tree diagrams) has traditionally been used by electrical engineers and computer scientists to illustrate the logical flow of data through a system. These diagrams can be created relatively quickly, prior to implementing complicated systems, and then manipulated to determine the optimum flow of data. The method can be used as both a generative exercise and an explanatory tool. Generally, this diagram is "emotion agnostic"—it focuses on actions and reactions of both people and the system itself. Building a Process diagram (see Fig. 6.13) is a form of synthesis that forces the designer to rationalize knowledge related to temporal flow and to constantly change his frame and consider the totality of behavior. (He is forced to consider not only an idea's core but also its periphery). A designer must consider the edge cases, describing what happens when less desirable behavior occurs with a product, service, or system.

A process diagram helps a designer move from data to information by contextualizing a discrete step in the context of the entire process, grounding a particular action or movement in relationship to other actions or movements. This forces the designer to evaluate human and system decision points and to consider what will happen when a person does something or behaves in a certain way. Additionally, once the

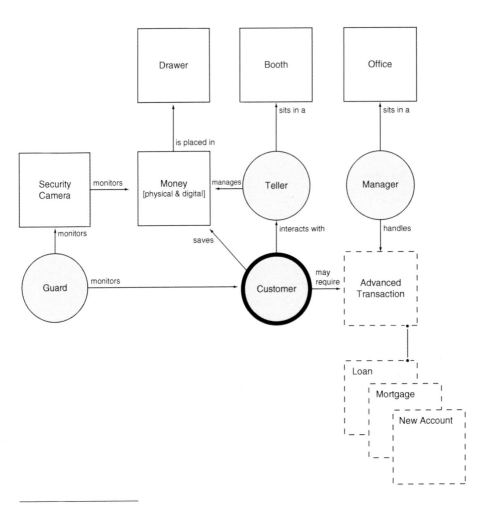

Figure 6.12
Flow diagram.

diagram has been created, it can be used for a number of later-phase activities. For example, in software design, the same diagram can form the beginning of a unified modeling language (UML) diagram suitable for authoring test cases, or it can be used to identify screen-based design criteria for the wireframing and visual design phases of software creation.

How to Apply This Method

Before creating a process diagram, a designer must have identified, through ethnography, the system's entities and their roles. An entity is typically a person, but it can be a group, a company, or another system or process. As an example, if I go into a bank and hand my money to a bank teller, we are three entities: me, the teller, and the bank.

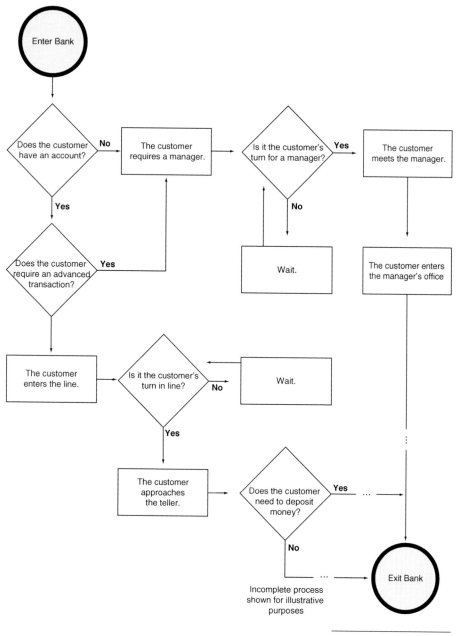

Enter Bank

Does the customer have an account? **No** → The customer requires a manager. → Is it the customer's turn for a manager? **Yes** → The customer meets the manager.

Yes (down from "Does the customer have an account?")

No (down from "Is it the customer's turn for a manager?") → Wait.

The customer enters the manager's office

Does the customer require an advanced transaction? **Yes** → (to "The customer requires a manager.")

Yes (down) → The customer enters the line. → Is it the customer's turn in line? **No** → Wait.

Yes (down) → The customer approaches the teller. → Does the customer need to deposit money? **Yes** ...

No (down) → ... → **Exit Bank**

Incomplete process shown for illustrative purposes

Figure 6.13
Process diagram.

Here are the steps to creating a process diagram:

1. List and prioritize the entities. Prioritizing, being subjective, can be approached from a variety of perspectives. For example, if the perspective is usability, the focus will be on end-user entities. If the

perspective is financial optimization, entities that generate or manage revenue may be deemed most important.

2. List the major processes, in the sequence in which they occur. A banking flow includes processes such as "enter the bank," "fill out the deposit slip," "stand in line," and "find out my account is overdrawn." Despite seemingly infinite processes that can occur in a bank, the designer can limit them by focusing on the ones most relevant to the prioritized entities listed earlier.

3. An ancillary benefit of the list is that it may—and likely will—identify gaps in knowledge that need further research. In the earlier banking example, while the designer begins to list what tellers do when they receive money from customers, she may quickly realize that she has no idea what happens to the money once it is placed in the till. So she must find out.

4. Additionally, note that the designer can select any level of detail upon which to create a process flow. Each process, such as "stand in line," has subprocesses that can also be mapped (like "approach the line," "take the last spot in line," "move forward," etc.).

5. Identify the start and end points for the first major process, tie them to the primary entity, and place them on a large sheet of paper ("customer enters the bank" and "customer exits the bank").

6. Map out the logic flow to connect the starting point and the ending point. Then ask logical questions to understand what a user will do next, and draw branching decision paths to show what happens at each decision point. In our banking example, a customer (entity) enters the bank (process). This has been defined as the starting state. Does he approach the line? (question). If yes, he stands in it. If no, he stands somewhere else in the lobby (branching path).

7. Continue mapping entity, process, question, and branching path throughout the entire described

Exposing the Magic of Design

experience until you reach the desired end state. This process typically requires many iterations, because it is difficult to anticipate the number of choices available to a person as he moves through even a simple system. Because of this difficulty, it is useful to work on large sheets of paper until you identify a core structure. Only then does it make sense to move to a computer for help in organizing and altering the presentation qualities of the diagram (such as making sure logical lines do not intersect).

8. Observe the process flow to find areas of complexity and needs for more research and to simplify processes that need it.

Although the diagram can be useful throughout the project, the act of creating it is most useful to the designer because it develops a strong mental representation of the boundaries of a complicated system. Essentially, she has built a mental model akin to actually performing the diagrammed functions. So although she may not have the tacit knowledge built from years of, say, working in a bank, the designer can now visualize changes to the system and more quickly and accurately predict how changes would affect different entities.

This type of diagram can be used to illustrate either the existing "problem" state or the potentially new "fixed" state. Both help to understand a design problem's structure and to reframe a static problem into a temporal one. In fact, creating a diagram for both the "before" and "after" states can be a powerful method of communicating design's positive impact from a systems perspective.

For Example:
The Flow through a Hunter Fan Thermostat
Contributed by Lauren Serota, frog design

Overview. Interaction with any object or service happens over time. Through new media (motion graphics, film, audio), we have learned how to manipulate time and what those manipulations mean. Although we cannot yet alter the cadence or speed of time as it pertains to real life, designers can alter and manipulate the order of actions within designed systems. Through this manipulation, we can create organization for experiences that feel more efficient, more playful, or more natural by matching the user's understanding and expectations. Task flow diagramming is a way of visualizing this *manipulation of time*.

Task flows are tools used in analyzing, designing, documenting, or managing a process or program in various fields. As designers, we typically use them to map out decisions and actions that users can take when using a physical product, a software application, or a service. Task flows can be very useful in breaking down anything that happens over time. They can be particularly useful for moving from user insights or a design brief into the initial stages of design. For example, if you are trying to understand opportunities for new tools and processes within a certain task or domain, an easy way to frame the context for that tool or process is to understand the sequence of events and the possible decision points around that task.

Consider a clothes dryer as an example. A typical, high-level task flow of using a clothes dryer goes like this:

open door > load clothes > close door > pick setting > press "start"

It provides a good example because each step and the physical embodiment of the object that facilitates each step can be clearly imagined. Imagine if, for some reason, you moved "pick setting" so the flow went like this:

pick setting > open door > load clothes > close door > press "start"

This is a possible flow with most clothes dryers. However, by framing the task this way, you can see the opportunity for design exploration aimed to support this flow if it were found to be more mechanically sound, useful, or enjoyable. A possible outcome of the revised flow could be the elimination of a "start" button, with the dryer instead starting when the user closes the door. Additionally, the alternative flow provides a means for comparison; with multiple styles of interaction to compare, a design team can debate the tradeoffs of each.

Tasks flows have both practical and conceptual applications. For example, they can be used practically as an outline for any kind of interaction (human to human, human to product, product to product) or as a guide to understanding decisions and actions that take place over time. That is how interaction and industrial designers typically use them. Task flows can also be used in understanding an ecosystem. When designing one part of a larger system, understanding how that part works and defining its role within the larger system is important and often inspirational for designers. Task flows are a good way of visually understanding this type of information.

Task flows can vary from the strikingly simple, such as the clothes dryer example mentioned earlier, to the overwhelmingly complex. This flexibility lends itself to the layering of useful information based on the intended goal of the task flow. For example, designer Austin Govella proposes including active versus passive indicators in a task flow to further enrich the usefulness of the diagram and using artifacts versus actions to communicate response (such as "see stop sign" [passive] and "stop car" [active] as two discrete pieces of the diagram). One could also imagine using a similar differentiation when designing a convergent physical/digital device, such as a cell phone or medical scanner, as indicators for physical versus software interactions. Govella also proposes indicating insights generated based on the task flow, as well as data output from particular steps to provide context for activities and design decisions.

As the experiences we have with products become richer and more complex, designers increasingly use task flows in the product development process to produce clear, functional visualizations for communicating with both the design team and with other stakeholders.

A Case Study

> Hunter Fan's Industrial Design team asked Lextant's User Experience team to learn and reframe how customers manage the comfort of their air at home, and translate that learning two ways: into a long-term strategy for the company and a short-term solution of a reimagined, easer-to-use programmable thermostat.

Our user research validated that programming thermostats are difficult to use. In fact, people had so much trouble programming their thermostats that they would eventually operate them manually, negating both the convenience and the environmental and economic benefits of automatic temperature adjustment. We found that current thermostats are difficult for three reasons: The thermostats unclearly disclose steps. They contain inconsistencies between the language used and way users think about their programming schedule. And the products lack clear direction. All of these difficulties are multiplied (and perhaps even created) by the technology constraints of typical programmable thermostats—the use of a segmented LCD display to reduce cost and extend battery life. This technology prohibits the screen from "refreshing," so each area of the screen can only be used for one graphic (for example, consider most inexpensive alarm clocks—these have multiple segments, allowing for each number to appear). Tasked with designing under this same display technology for our product, we relied heavily on task flows to help understand logical task sequence, and control and indicator positions.

Current programmable thermostats ask users to program temperatures based on four specific time periods: waking up, leaving the house, returning home, and going to sleep. But our research also indicated that people tend to think about their day in terms of their context—when they are "at home," "away," and "sleeping," rather than as specific points in time. Rather than saying, "I want it to be this temperature at (exact time)," the research team heard customers saying things like "I want it to be this temperature while I am (away, at work), between (time) and (time)."

The team used task flows to transform the programming steps from the traditional model (setting four distinct time points, per the beginning of each "period") to the new model, and to address some of the difficulties stated previously. We broke the current programming steps into individual actions and possible outcomes so we could understand where and how a new programming flow could shift, add, or remove steps. We looked at both programming the thermostat for the first time and going back to adjust specific times and temperatures. The user's goal was to set the thermostat for different temperatures at different times for different days of the week. The goal of our initial design was to establish time periods and temperatures as different "rooms" the user would enter, which focused more on adjusting the thermostat (which happens more often than the initial programming).

By diagramming the sequence of events, we could begin to establish an information architecture for the interface. The task flow shown in Figure 6.14 was used as an underlay for a more sophisticated task flow, including discrete individual actions (using +/− to adjust hours and minutes) and all possible paths via "accelerators" (hidden shortcuts). Task flows can also seamlessly transform into information architecture for interactive experiences or discrete steps and actions for physical ones. They can simultaneously act as tools for understanding and as guidelines for initial screen layout (wireframing). In this case, the thermostat has both hardware and software components, so task flows helped to establish the logical locations for physical controls and for indications of next steps in a task.

Users evaluated our prototype of the initial interface that emerged from the first task flow. As they went through our outlined tasks (including "set the times and temperatures on this thermostat to match your schedule"), we noticed opportunities for improvement in the programming area. Users easily understood the new concept of the "away," "asleep," and "at home" periods. However, they experienced three areas of difficulty:

1. Identifying both the time and temperature desired for each period

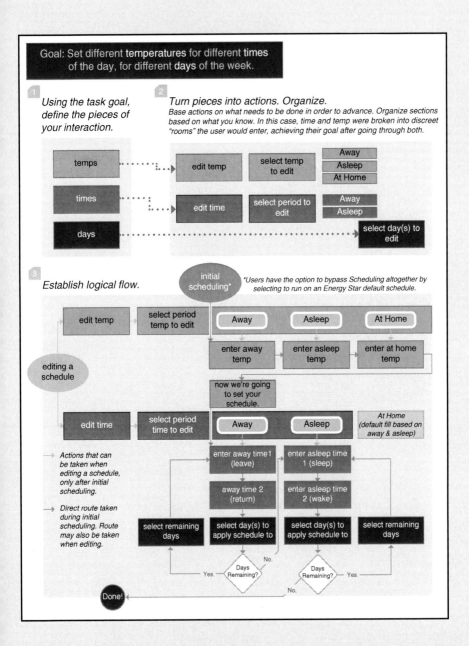

Figure 6.14
Task flow: early stage.

2. Programming all three periods under temperature, but only two under time

3. Selecting the specific days to program as a discrete and separate step, which is a convention established by most current programmable thermostats.

Exposing the Magic of Design

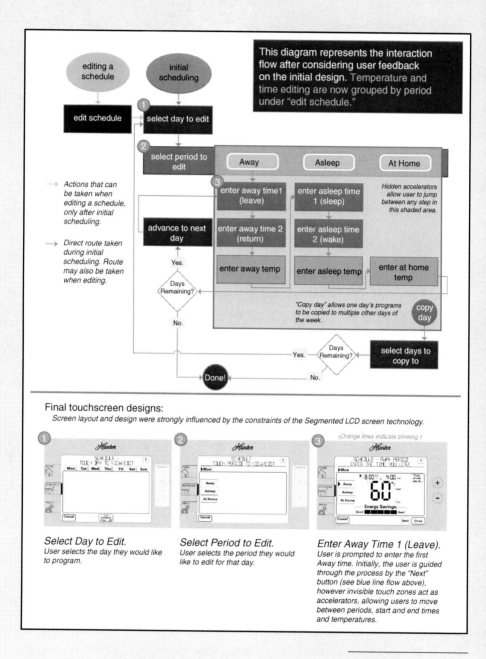

This diagram represents the interaction flow after considering user feedback on the initial design. Temperature and time editing are now grouped by period under "edit schedule."

Final touchscreen designs:
Screen layout and design were strongly influenced by the constraints of the Segmented LCD screen technology.

Select Day to Edit.
User selects the day they would like to program.

Select Period to Edit.
User selects the period they would like to edit for that day.

Enter Away Time 1 (Leave).
User is prompted to enter the first Away time. Initially, the user is guided through the process by the "Next" button (see blue line flow above), however invisible touch zones act as accelerators, allowing users to move between periods, start and end times and temperatures.

Figure 6.15
Task flow: later stage.

To address the first two areas, we combined the times and temperatures. We also built in hidden shortcuts to program adjustment that did not interfere with the user's ability to complete the task in a linear manner (see Fig. 6.15).

This overcame the drawbacks of the segmented LCD technology by navigating users through steps that needed to be linear (screens 1 and 2 in Fig. 6.15), while allowing for more flexible navigation in areas where more information was helpful in providing context (see screen 3 in Fig. 6.15). The third area was resolved by moving "select day to edit" to the beginning of the flow.

Figure 6.15 illustrates definite connections between the task flow and the final design of the thermostat interface. Task flows allowed the team to frame the best way to progressively disclose information and actions to the user.

As the design progressed, task flows were important for articulating the reasons for changes in both the product's architecture and visual design. They allowed our team to explain to each client stakeholder why the control locations and engagements were key to the user's overall ease of use and cohesion of the experience. Task flows also allowed the stakeholders to ask questions and direct feedback at very specific points in the design, making their input easy for us to act on. Although task flows served as important communication tools between our two teams, they did not replace a fully functioning interactive prototype, which illustrated the task flow and possible deviations in a more comprehensive and realistic manner.

About the Author. Lauren Serota is an interaction designer at frog design, and was previously a design researcher with Lextant in Columbus, Ohio. Before earning a bachelor's degree in industrial design from the Savannah College of Art and Design, Lauren spent her formative years in public relations and promotions for the electronic music industry. She developed her own cultural anthropology curriculum while at SCAD, and she continues to seek out information and perspectives that challenge and reinterpret norms. Lauren has spent time as an industrial design recruiter, interaction designer, and trend researcher.

Chapter 7
Methods for Building an Experience Framework

By now, you have created a sense of informative meaning out of the gathered data through the various organizational and structural arrangements described in Chapter 6. This data may be meaningful on a pragmatic level—it will likely be more usable and useful (both to designers and to users) than the raw words, photos, videos, or artifacts captured during research. Additionally, these new visual descriptions will be easier to present to stakeholders and to others involved in the development of products. However, humanity requires a sense of emotional resonance. Usability and usefulness do not recognize the subjective and rich experience of the human condition. Only when the content is related to people and to the temporal qualities of behavior do the ideas really begin to work.

A great deal has been written about the nature of human experience. Philosopher and author John Dewey has made vivid and critical connections between experience, art, and education. Yet even Dewey has acknowledged that it is nearly impossible to predict an experience accurately for the simple fact that people "complete" the experience, and people are different. Dewey explains that "experience does not go on simply inside a person . . . Every genuine experience has an active side which changes in some degree the objective conditions under which experiences are had" (Dewey, 1997). Designers have increasingly embraced the idea of experience design. Despite the difficulty of structuring a repeatable and predictable experience, they can *approximate* the human behavior associated with time-based design by creating an experience framework: a scaffold that contains the elements of experience but allows for individual difference and expression.

Three techniques can be used to build the experience framework. All of these techniques emphasize the unique, rich, and complicated facets of human life.

Telling a Story

Traditional industrial and graphic designers focus on the creation of a static, two- or three-dimensional artifact. Designers focus on form, function, and comfort or aesthetics, but placing an immediate focus on an artifact implicitly places value on the creation rather than the user of the creation. When creating a scaffold for experiences, you should shift the emphasis to acknowledging and understanding the user and her activities, feelings, and desires *over time*. An easy, accessible, and relatively common way of articulating these issues is the story. Storytelling, as old as language, allows an individual to paint an immersive scene and environment.

The components of a successful story are subject to debate, and the ability of one to actually *tell* a successful story requires a particular talent. However, you can include some fairly common components to make your story believable and easy to understand.

The story should include a plot, characters, a setting, a climax, and an ending. These components can be presented precisely and vividly to encourage an awareness of sensory detail. The narrative should also include some underlying point. The story can be used to illustrate the problem with an existing design or situation, or to better show how an artifact, service, or system can be integrated into human day-to-day life. In either case, it should serve to humanize the design opportunity. The structure and emphasis of the story should be on people and goals rather than on technology or engineering. This will shift the emphasis from an artifact to an experience, helping to create an experience framework to move from information to knowledge.

Changing the Scale

Design problems exist in a context. You can always "zoom out" and articulate how the context fits into the larger framework of human existence. Similarly, you can "zoom in" to identify details or nuances that are otherwise glossed over. By zooming in, out, or even across, ideation and design investigation can be extended with regard to geography, time, or even subject matter.

For example, when designing a coffee maker (usually considered a physical artifact), you can zoom out to examine the context of use: the countertop, the kitchen, a residential location, or even a particular city or geographic region. This shift allows you to find new points of contact with consumers, new ideas for form and function, and new cultural requirements that can affect comprehension and desirability. Similarly, you can zoom in to examine the details—the details of freshly ground coffee, the grinder, or even of the coffee tree itself where the beans are grown. This forces you into a position of conceptual thinking rather than pragmatic thought. You can also "zoom across" a problem in space and time, thinking back through the roots of the issue and projecting forward in time to see how a design might evolve. This semantic zoom can help identify opportunities throughout the entire problem space, rather than simply at the typical point of emphasis (the product, or artifact). As an example, you may start by exploring the "experience of use," then backtrack to consider how the product was purchased, how it was shipped to the store, or even how it was produced.

Shifting the Placements

Another way to understand experience is to understand the placements that support design, then to consciously and very explicitly shift these placements. Theorist Richard Buchanan describes placements in his text *Wicked Problems in Design Thinking*; he explains that ". . . signs, things, actions and thoughts are not only interconnected, they also interpenetrate and merge in contemporary design thinking with surprising consequences for innovation . . ." (Buchanan, 1996). These are the placements, the organizational schema Buchanan has established to describe all designed elements. Signs and things represent symbolic, visual, and physical communication. Actions imply the idea of activities or services, whereas thoughts can represent complex systems. Buchanan goes on to explain that the real value of the placements comes when a designed opportunity moves between points in the placement framework: ". . . innovation comes when the initial selection is repositioned at another point in the framework, raising new questions and ideas" (Buchanan, 1996).

A coffee maker is a *thing*. You may approach the problem of coffee making in the home by exploring various physical shapes and functions that can make coffee making more usable, useful, or desirable. But more interesting—even innovative—results occur when considering that the output of design activities may be signs, actions, or thoughts. What if you come up with a coffee provider that is part of a system instead of a physical artifact? Perhaps you identify an opportunity for a coffee delivery service or a new type of digital coffee-related interaction (a consumer could control the exact taste of the beans through digital controls). These interesting relationships are forged by viewing the old design opportunity in a new light: by shifting the placements of the design framework. Ultimately, this shift forces you to reconceptualize your design hypotheses and helps to create a framework for experiences (see Fig. 7.1).

The methods presented in this section—methods of moving from information to knowledge—help build an experience framework, in which people can have positive and reflective experiences. Each of the methods and examples emphasizes a form of storytelling that shifts perspective and that positions you as a storyteller who can create time-based narratives related to experiences.

Method: Concept Mapping

A concept map is a graphical method for organizing and representing knowledge. Joseph Novak and Alberto Cañas—arguably the method's creators—describe how a concept map supports meaningful learning through associative and hierarchical connection forming. It "serves as a kind of template or scaffold to help to organize knowledge and to structure it, even though the structure must be built up piece by piece with small units of interacting concept and propositional frameworks" (Novak & Cañas, 2006).

The Placements

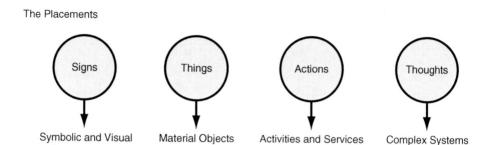

Shifting The Placements

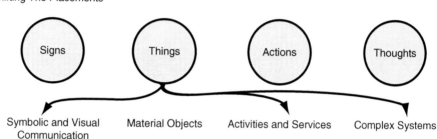

Essentially, the map can be thought of as a picture of understanding (Kolko, 2007a). It is a formal representation of a mental model. And a mental model "represents a possibility, or, to be precise, the structure and content of the model capture what is common to the different ways in which the possibilities could occur . . . when you are forced to try to hold in mind several models of possibilities, the task is difficult" (Johnson-Laird, 2006). The concept map itself represents the creator's mental model of a concept, but it also informs and shapes that mental model as it allows designers to see both the holistic scale of the concept and also critical details within the concept. Because it affords action-based understanding at both a gross and fine level, both its creation and its usage become tools for sensemaking.

Novak and Cañas describe three core learning approaches that are emphasized during concept-map creation:

1. *The designer must possess prior knowledge.* This implies that the designer actually conducted the research to generate the data that are being synthesized.

Figure 7.1
The placements of the design framework.

A discovery process, frequently considered a "waste of time" because of the lack of actionable output, is actually critical for developing tacit knowledge upon which this technique can build.

2. *The material must be presented in language that relates to the designer's prior knowledge.* Because the "learner" is the designer in this case, he selects the language representations used on the map and he embeds in the map various elements that make sense to him, based on his prior knowledge. As he integrates new ideas into the map, they build upon a recognized foundation of previous experience.

3. *The designer must choose to learn meaningfully.* This implies that he must see the value in synthesis methods and in concept mapping because a more automatic or procedural approach to this method will act as a more rote, less effective method of learning. (Novak & Cañas, 2006)

Generally, a concept map links elements to one another. It creates a hierarchical relationship between elements and allows for cross-linking between seemingly disparate elements. *Specifically*, a concept map will form connections between entities (nouns) by describing action-oriented relationships (verbs). The connections are literal, because a noun will have a line drawn from it to a verb that makes sense in the given context of a design problem. The map provides a visual way to understand relationships through literal connections as well as through proximity, size, shape, and scale.

As an artifact, the map is intended to illustrate relationships. The map can be used to communicate these relationships to other members of the design team or to a client. More important, however, the act of creating the map is generative and is a way to produce critical knowledge. You must make subjective value judgments in both selecting the items to include on the map and in indicating the relative strength of the relationships between items. In this way, you are actually learning as you make this artifact. In fact, the artifact is almost incidental when compared to the amount of highly specific domain knowledge you will gain while making the artifact.

How to Apply This Method

You can produce a concept map by following these steps:

1. First, begin to identify the words that make up the core taxonomy. A taxonomy is a classification of words, often arranged in a hierarchy of importance. In this case, the words that matter are the nouns and verbs of the problem space—the people, places, systems, artifacts, organizations, actions, processes, methods, and other entities and activities. List all of the words on index cards or post-it notes, one word per card. See Figure 7.2 for a simple and unranked taxonomy in an (oversimplified) example of teeth brushing.

2. Now create the sense of order and hierarchy in the taxonomy by rearranging the cards. Elements are deemed to be more or less important than one another, and they are moved to illustrate this importance. Those that are identified as a subset of a larger element are indented and placed lower to illustrate this relationship. New elements are added at this stage as appropriate.

This prioritization forces the design team to make value judgments, arguing for or against a particular position, about each item based on each team member's understanding of the problem space. This activity works best in teams of two or three people, as the discussion generates useful information about the dataset. The teeth-brushing taxonomy may be prioritized as shown in Figure 7.3.

3. Begin to outline the structure of the map. Using the most important elements (or the elements farthest to the left in the indented hierarchy), begin to draw, on a large sheet of paper, circles to illustrate each entity. Connect the circles with lines to illustrate relationships between the elements.

Figure 7.2
Simple and unranked taxonomy in an example of teeth brushing.

Figure 7.3
A prioritized teeth-brushing taxonomy.

4. Continue to fill in the rest of the map. It will begin to create small sentence fragments of meaning, such as "teeth can become clean by using a scrubbing motion" (see Fig. 7.4). This illustrates the generative and subtly abductive nature of the map, because you may have no deductive or inductive way of knowing that teeth can become clean by using a scrubbing motion.

5. Complete the map. This will likely take several tries, as you begin to understand which elements have more nodes, and how deep connections between nodes may not have been obvious in your static list of terms.

6. Use a visual design tool such as Adobe Illustrator to clean up the mess and make the contents easier to understand. Selectively add color to create another level of hierarchical organization.

During synthesis, you can use the concept-mapping method as described earlier to organize and understand a topic and to produce a model of that understanding. The model will continue to change and evolve as the design does.

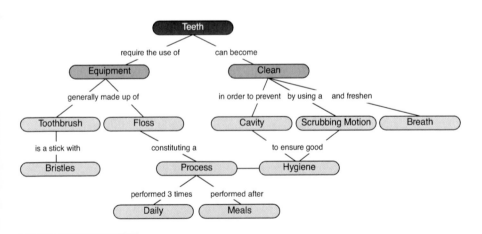

Figure 7.4
Filling in the map with small sentence fragments.

For Example:
Using Concept Maps in Product Development
Contributed by Hugh Dubberly, Dubberly Design Office

A concept map is a collection of terms related to a main idea. Links between terms form a structure—something like an outline, but with some branches connected. Labeling a link with a verb creates a noun-verb-noun chain that can be read as a sentence. Thus, concept maps present a series of propositions related to each other and a main idea.

Mapping a content domain—creating a concept map—is an effective way to understand a domain. Sharing a concept map with project stakeholders is an effective way to identify errors in understanding and reach consensus on content definition, structure, and boundaries. Mapping a content domain is a good way to prepare for designing or redesigning a content-rich Web site, application, or service.

The Benefits of Concept Mapping

Deepening Understanding. We developed a concept map of the Java programming language as a way to understand Java. The map helped us prepare to redesign and relaunch Sun's main Web site for Java developers, www.java.sun.com. Concept mapping was one of many tools we used in the design process, which also included auditing the existing site, reviewing site traffic logs, and interviewing Java developers. The main questions we faced were these: How should we organize www.java.sun.com? What should the information architecture be? Answering these questions was not trivial because the site contained more than 110,000 pages. It could not be reorganized by simply reading a few pages and moving them around. What we needed was a deep understanding of Java—what it is, how it is used, how it changes, and why it matters.

The trouble was as follows: We knew little about Java except that it was a programming language that runs in many environments. We developed the Java concept map so we could learn what we needed to know. The knowledge we gained making the map enabled us to propose revisions to the site's information architecture with confidence. The knowledge also helped us to back up our proposal with reasoning built on a firm foundation—a definition of the content domain (i.e., the Java concept map) already accepted by the client and his many internal constituents.

Building Trust. Like any large corporate project, the redesign of www.java.sun.com encountered political issues. First, it was a visible project in a decentralized company. That meant the project had a lot of vocal stakeholders. In addition, www.java.sun.com was managed by Sun's Developer Relations Group, which had recently been formed by consolidating several previously separate departments. Not everyone was happy about the new organization.

As we began to meet internal stakeholders, we encountered considerable skepticism about the site redesign project and our ability to execute it. Developing the concept map became a way to engage known stakeholders, discover new ones, and build trust.

We interviewed a series of Sun employees involved with both Java and www.java.sun.com. We began with a small group of employees, who in turn suggested others. Eventually the number of employee interviews exceeded 50. We also asked the stakeholders to review the concept map as we developed it.

At a project meeting a few weeks into the process, one of the key stakeholders reviewed the map and said, "Not bad. It looks like you're ready to meet the Java Distinguished Engineers." Before that, no one had mentioned these high priests of Java; they turned out to be a powerful constituency. The map helped us find them and gave us *entree*—both permission to meet and something to discuss. Those meetings went well; the Distinguished Engineers were intrigued by the map. (It is not often that someone turns up with a map of your baby.) We also entered the discussions with more credibility than we had at the start of the project, because we had clearly done a lot of homework to make the map. Our organizational efforts paid off, because Sun introduced the Distinguished Engineers only when we were prepared to meet them.

The most important benefit of the map, though, was that we were able to discuss the structure of Java and ensure that we understood it, rather than discuss a menu system or page layout, which might have conflated issues—the structure of Java, the site information architecture, and the appearance of the navigation interface.

By separating content from expression—by mapping—we were able to establish relationships and build credibility and trust before proposing changes to the client's baby, the Java Web site.

Other Uses. Although the main goal of the concept map was to help the design team understand Java so that we could reorganize www.java.sun.com, it soon became clear that the map might have wider uses. Our working version of the map looked like a sketch, which reflected the constant changes we were making. (It was messy.) The sketch form invites comments where a more polished form may inhibit them.

When we reached consensus on the content, we formalized the map's appearance. Eventually, the map went through two printings and was distributed to more than 25,000 Java developers. We also created an interactive map, which is still available online at http://www.dubberly.com/concept-maps/java-technology.html

The Process of Concept Mapping. At the beginning of the www.java.sun.com redesign project, we asked to see Sun's Java models. Although we were unable to find detailed models, we did find slides from marketing presentations—"marketectures," simplified versions of technical architectures. One of these marketectures depicted Java as the Parthenon; three steps supported a few columns capped by an architrave and a pediment. This model included less than a dozen elements. It became our starting point.

Set Goals. Setting goals is the key to managing. Rick Robinson points out that all research should begin with a clear goal, what he calls a "hunt statement." Likewise, mapping should begin with a clear goal.

We set six goals for the Java concept map:

1. Develop an understanding of Java shared among the www.java.sun.com redesign project's stakeholders.

2. Inform both the logical organization of www.java.sun.com and its integration with other sites.

3. Develop a framework by which changes to Java can be understood.

4. Open a dialog with senior Java stakeholders.

5. Provide an overview of Java to people familiar with computing but unfamiliar with Java.

6. Develop a map that an average Java programmer would consider accurate.

Identify Terms. The first step in developing a concept map is to identify terms that could be included. In this phase, the goal is to quickly explore the domain. Write down whatever you find or think of. Editing comes later.

Our first list of terms came from the team's own experience, from glossaries of Java terms, and from the indices of books on Java.

We kept our list of terms in a spreadsheet. We printed each term on a label and affixed the label to a colored "sticky," so that it could be moved and grouped later. We then placed the stickies on a 4-by-8-foot foam-core board, so that we could move the whole group around the office easily (see Fig. 7.5). Our initial list included roughly 400 terms.

Prioritize Terms. We prioritized the terms, creating more manageable clusters:

- 11 first priority

- 45 second priority

- 157 third priority

- 136 fourth priority

- 51 fifth priority

Triage is a similar strategy. Which terms are critical? Which terms can we deal with later? And which terms are not relevant?

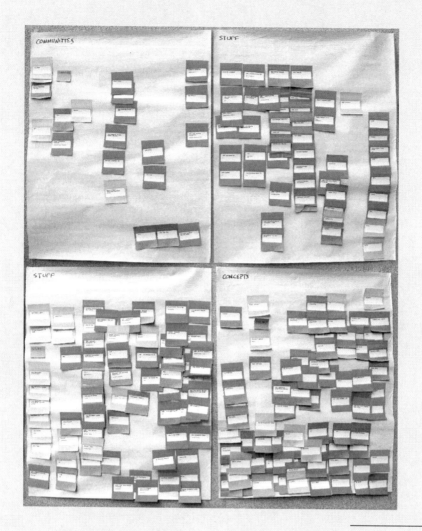

Figure 7.5
Example of stickies on a foam-core board, which enabled easy movement around the office.

Define Terms. We defined each first-, second-, and third-level term, adding definitions to the spreadsheet. The list of definitions served as a foundation for later work. In discussions with reviewers, the definitions allowed the team to focus on individual words, without referring to the map. The list of definitions was particularly useful in conversations with reviewers who did not understand that map, especially when they reviewed early versions. Two hundred and five definitions were collected from eight sources.

Organize Terms. We organized the first-, second-, and third-priority lists into a single outline. We experimented with several variations. For the most part, category titles in the outline were first-priority terms.

Test Armatures. When the number of terms in a concept map exceeds nine or ten, introducing levels or hierarchy may make reading easier. Large concepts maps (more than 50 terms) are almost impenetrable without attention to both semantic and visual hierarchy.

We like to organize large concepts maps around an "armature," a primary sentence or two. A good place to start is with a horizontal sentence placing the main concept in a context; then add a vertical sentence defining the concept. Other terms link off the armature.

An armature should include the terms most fundamental to the concept being mapped. These fundamental terms and relationships serve as the backbone for the rest of the map, providing structure and hierarchy. The armature is often a starting point for readers (see Fig. 7.6).

We experimented with several armatures. The client and the design team chose the armature with the most meaningful relationships and the one that provided both physical and logical space for the rest of the terms.

Add Terms. We added second- and third-priority terms. New terms suggested changes to the armature.

Review and Revise. Once we had an armature fleshed out with secondary terms, we reviewed the map with the client

Figure 7.6
Example of an armature.

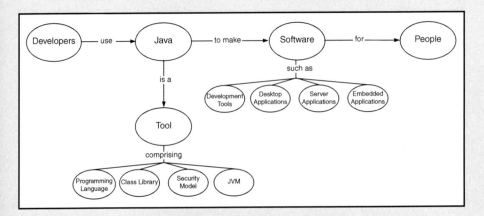

Exposing the Magic of Design

and a small group of Java experts. They suggested additional reviewers. From this early stage, reviews were ongoing. We continued to interview stakeholders while we developed the concept map, asking them to review and comment on the current version (see Fig. 7.7).

Reviews took place in one-on-one interviews, on the phone, or via e-mail. We sent drafts of the map to groups within Sun. We also posted large printed copies in high-traffic areas at Sun; reviewers wrote directly on the map or attached yellow stickies. Marked-up maps were returned to us. Several people reviewed the map multiple times. Thirty-six people reviewed the map in one-on-one interviews. Ten people provided feedback via posted maps.

Subdivide Large Maps. As we added terms, the map became unwieldy and difficult to consider as a whole. So we divided the map into logical sections (see Fig. 7.8).

Subdividing the map increased efficiency. We distributed sections to team members, who refined their sections simultaneously. They added terms, modified relationships, and, in some cases, created secondary armatures. We reassembled the sections around a refined armature, paying special attention to relationships between the sections. At its largest, the map measured 3 x 8 feet (see Fig. 7.9).

Refine the Typography. The team adopted Sun Sans as the primary typeface, conforming to Sun's corporate identity standards. Early sketches produced some new typographic devices that were eventually applied to the map. One device was a sort of footnote or hypertext link, which allowed us to indicate more relationships without drawing more long lines across the map (see Fig. 7.10).

Refining appearance required seven complete revisions.

Check Again. Throughout the project, we worked with a copy editor. She checked each comprehensive revision for spelling, grammar, and sense. Sun's legal and trademark department also reviewed the map several times, as did an attorney working for our client's department and two subject-matter experts.

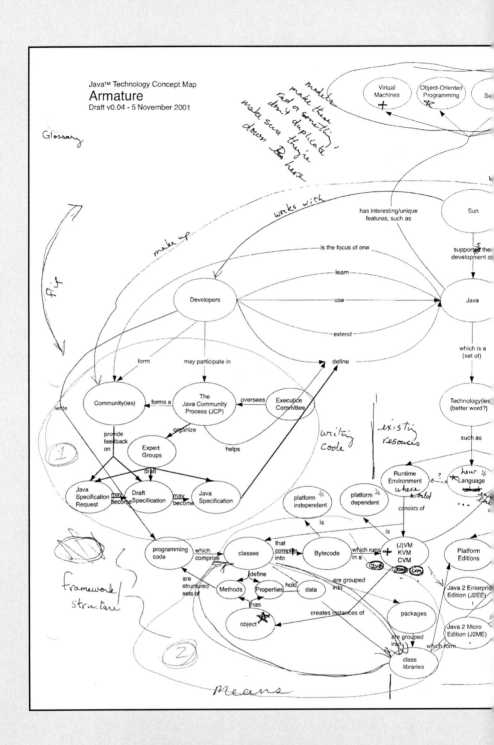

Exposing the Magic of Design

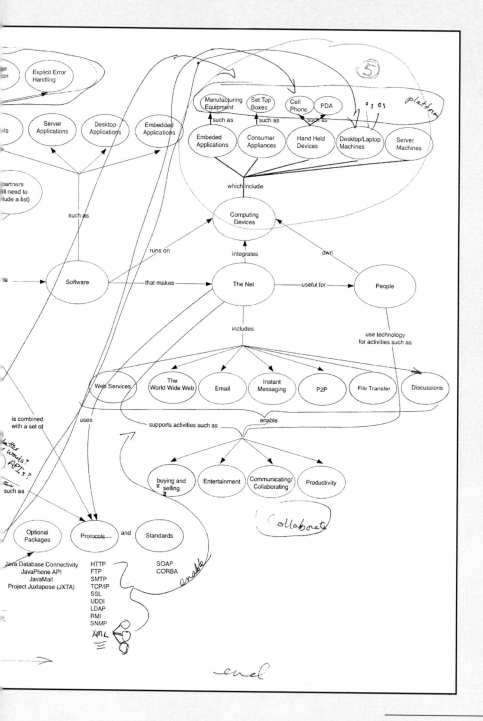

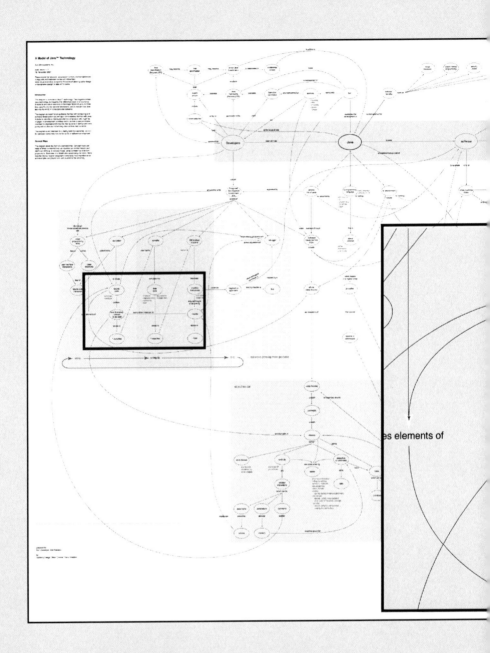

es elements of

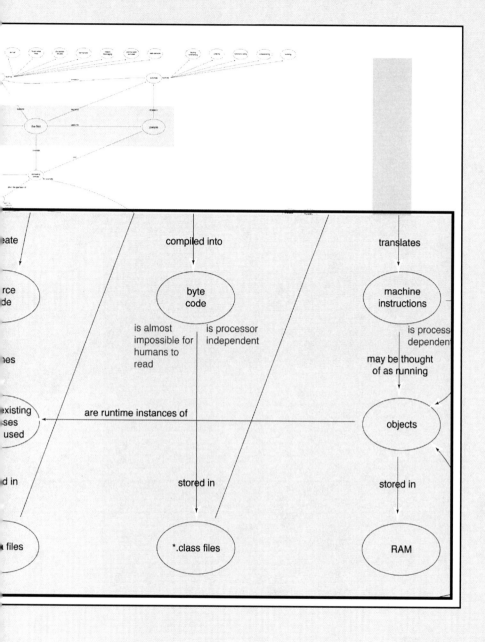

eate

compiled into

translates

rce
de

byte
code

machine
instructions

is almost
impossible for
humans to
read

is processor
independent

is process
dependen

may be thought
of as running

es

existing
ses
used

are runtime instances of

objects

d in

stored in

stored in

files

*.class files

RAM

Figure 7.8
Concept map divided into
sections.

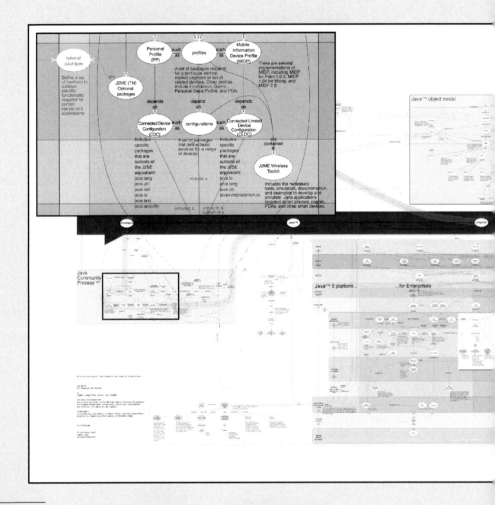

Figure 7.9
The subdivided
concept map.

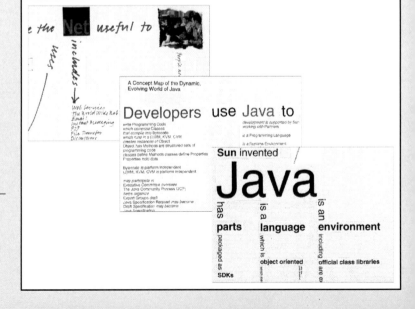

Figure 7.10
Concept map using
hypertext link, which
allows for indicating more
relationships without
drawing additional lines
on the map.

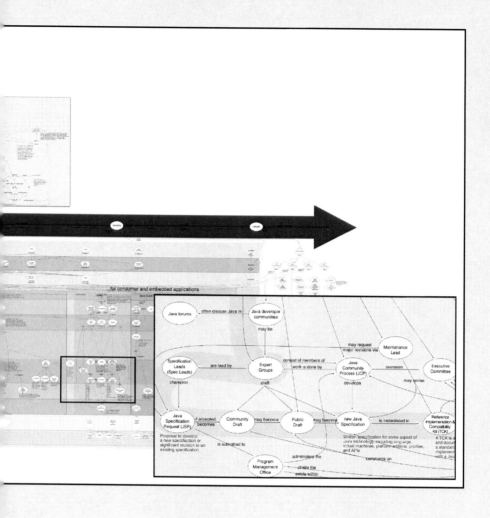

At the end, Sun's marketing department asked for a few visual changes—and we faced a nerve-racking few days when a senior manager questioned whether the map contained too much proprietary information. Luckily we were able to show that the information was already available on www.java.sun. com. The map went through a total of 53 numbered variations (see Fig. 7.11).

Print and Distribute. The map was physically printed, and Sun initially distributed the map at the JavaOne conference in Japan.

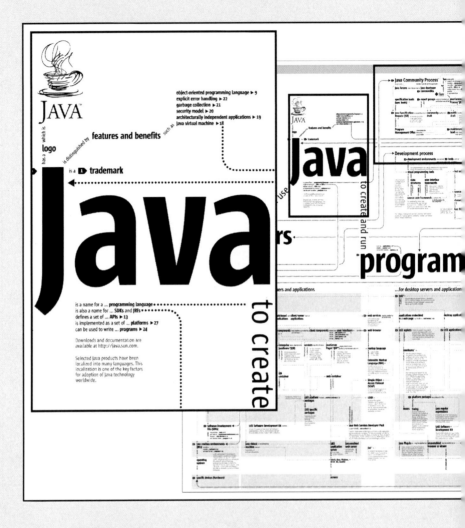

Figure 7.11
Variations of concept map.

Project Stats. The final map contains the following:

- 235 terms

- 425 links (relationships)

- 110 descriptions

We began the map in October 2000 and delivered printed copies in September 2001. The process required the following:

- 49 weeks

- more than 50 interviews

Exposing the Magic of Design

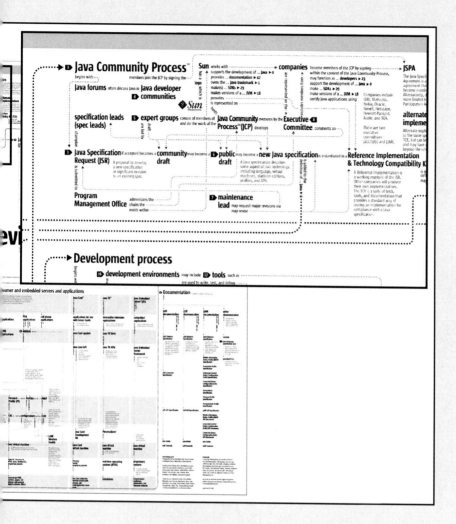

- more than100 meetings

- more than 2,000 emails

The team that created the map included the following individuals:

- Audrey Crane, project manager, interviewer, researcher, mapper

- Paul Devine, content expert, mapper

- Hugh Dubberly, interviewer, mapper

- Jim Faris, mapper, graphic designer

- Paul Pangaro, our client

- Harry Saddler, content expert, mapper, graphic designer

- Ylva Wickberg, interaction programmer

More Information. For more on concept mapping, read Gowin and Novak's *Learning How to Learn.*

For more on teaching concept mapping, read Dubberly's *The Baseball Project: A Step-by-step Approach to Introducing Information Architecture,* available at http://www.dubberly.com/articles/the-baseball-projects.html

---

About the Author. Hugh Dubberly is a design planner and teacher. At Apple Computer in the late 80s and early 90s, Hugh managed cross-functional design teams and later managed creative services for the entire company. While at Apple, he cocreated a technology-forecast film called *Knowledge Navigator,* which presaged the appearance of the Internet in a portable digital device. Also while at Apple, he served at Art Center College of Design in Pasadena as the first and founding chairman of the computer graphics department. Intrigued by what the publishing industry would look like on the Internet, he next became director of interface design for Times Mirror. This led him to Netscape, where he became vice president of design and managed groups responsible for the design, engineering, and production of Netscape's Web portal. Hugh graduated from Rhode Island School of Design with a bachelor of fine arts degree in graphic design; he earned a master of fine arts degree in graphic design from Yale University.

Method: Forced Semantic Zoom ("Ecosystem Mapping")

In 1977, Charles and Ray Eames created a film called *Powers of Ten*. Through a constant perspective change, the film first explores the very large (the solar system) and then the very small (atoms and molecules). The viewer can frame the subject matter in relationship to a "known" form (the human body), and as the visualization zooms in and out, the viewer can draw conclusions about the relative size of otherwise unthinkable concepts (just how big is 10 to the 24th power, anyway?). The film is a literal journey through geography, space, and time, but it also illustrates a more important and subtle principle: Changing the scale of a problem illustrates new problems, issues, and opportunities, and it allows the designer to recontextualize the problem (Eames & Eames, 1978).

This form of literal zooming—of looking at things from a closer perspective, or from farther away—matches closely the semantic zooming that Charles Eames referenced as being critical to his creative design work. As previously discussed, you can "zoom in" on a problem space to focus on the details, uncovering new problems and design opportunities in the nuances of a specific area of the problem. Additionally, you can "zoom out" to look at the container of the problem—often a larger problem itself—and to better understand contextual cues and inherited problem details. This zooming describes the ideas of inheritance and hierarchy, because a single problem may contain other problems (inheritance), and problems are hierarchically related. This describes a new way of thinking about a problem space, and one that can be used during design synthesis to develop a more refined sense of purpose or a larger frame of reference. Consider that you can look "above" the problem—zooming out and broadening a perspective—to understand the larger context of the problem. This illustrates parallel problems or opportunities, describes the relative size of the problem space, and helps to contextualize the language of the problem space in a larger, broader vernacular.

As an example, imagine that you are creating a Web site for a cellular telephone service provider, Texas Telephone.

This Web site will let users buy phones, plans, and accessories. At the moment, you are tackling a particularly difficult problem: how to structure the services area of the site, where a user can choose from various options:

Some services will work only on phones with a large screen.

Some services will work only with family plans, where two or more users share a plan.

Some services are free, but have a per-use charge.

Other services cost a flat rate per month.

Still others are free only with a monthly data plan.

In your mind, the problem space likely starts as a mess, but through synthesis, it becomes more clearly defined. You can draw a map of specific services, with lines connecting them to plans. You might make a visual list of relationships, and you can start to see patterns emerging in the form of taxonomy. This taxonomy shows types of services (free, pay per month, pay per use), service penalties (overages, activations), plans (family, individual), content (data, voice), and phones (smart, feature).

At this point, you can continue to synthesize in the defined problem space. However, by "zooming out" a bit, you can start to see that you have ignored issues of customer service, bill payment, or the checkout process. These topics—related, but broader than services—clearly have an impact on services from the user's perspective.

By contrast, zooming in and narrowing perspective begins to identify actual problem constraints (as compared to theoretical domain constraints). Zooming in on the services mentioned earlier, you may start by grouping the services by genre (entertainment services, GPS and location services, communication services, etc.). This helps to make sense of the problem at one hierarchical level of detail, but it ignores the contents of the specific group. Now you can zoom in to look more closely at a single genre—communications services include text messaging, picture messaging, and one-to-one paging. Now you can zoom in to text messaging, finding that each message costs five cents and can be only 160 characters. *Now* you can zoom in to the character limit, to understand the historical motivation behind the limitation and the relationship to single-directional pagers.

You could follow this semantic zoom much more deeply or into the problem or much farther out from it, but neither would lead to a practical solution. Zooming way out would show that telephones exist in a culture and are deeply related to the nuanced relationships between people. You could explore those social relationships and digital technology's influence on them. This perspective has design implications, but they are too broad to be practical in the context of a real problem. Or you could zoom in far enough to explore various user-interface mechanisms for selecting a particular service, debating the pros and cons of a dropdown box as compared to a slider. During synthesis, this might be thought of as "stuck in the weeds," as this level of detail is inappropriate when attempting to tame the larger frame of complexity.

A forced semantic zoom, or ecosystem map, is most useful when used with a concept map, described earlier, and when a zoom is limited to only three levels up or down. As you have seen, a concept map represents a sketch of your understanding at a given time. You can force a semantic zoom into any area of the map and can act on this zoom by filling in the new noun and verb relationships that show up at that new level. These new relationships will be more tactical and will often describe more localized and nuanced behavior. Additionally, you can force a semantic zoom out of the map, and you can draw the noun and verb relationships around the perimeter (showing entities and actions that exist on a broader, more strategic level).

Crispin Porter + Bogusky, an advertising agency known for, among other things, the "Sacrifice your Facebook friends for a Whopper" campaign, uses ecosystem maps as a way of abstractly diagramming the anatomy of a new advertising program. These maps are a tool for understanding context—and this type of diagram fleshes out the relationship of the elements, their roles with one another, and overviews the essential campaign strategy.

Jordan Clayton-Hall, a designer at Crispin Porter + Bogusky, explains:

the process of making these maps forces us to think through all the campaign pieces and how they all fit together. This helps us understand the strategic/engagement environment. Through this process, we can look at where we are going, what we could be doing, and analyze any potential gaps we may need to bring to life. This helps us understand the conditions surrounding our idea and what is needed for it to grow in a complicated campaign system. More and more campaign initiatives we start now involve the visual planning of the campaign strategy, its functions, and how all the different parts of the system interact with the central idea as a whole. In many ways these maps are the campaign visualized. Once we have figured out what the key story is about, we can then lay out what needs to be articulated in the map.
(J. Clayton-Hall, personal communication)

How to Apply This Method

Applying the semantic zoom to a concept map requires follow-through. It can seem redundant and time consuming to reproduce elements of the concept map at each zoom level, yet capturing both the redundant conceptual "container" and the changing content within is entirely the point of this method. Here are the steps:

1. Sketch a concept map, a picture of understanding. When viewed from the perspective "zooming," the map actually describes *one* picture of understanding at a particular distance from the information (zoom level 0). Label the concept map "zoom level 0." Commonly, zoom level 0 describes a single product to be designed. It references major areas of functionality, user benefits, marketing attributes, and technology. It also likely describes the people who will interact with the product in some way (see Fig. 7.12).

2. Zoom out and redraw the concept map from that perspective, as though you had pulled the camera farther back while shooting the map. Individual

Exposing the Magic of Design

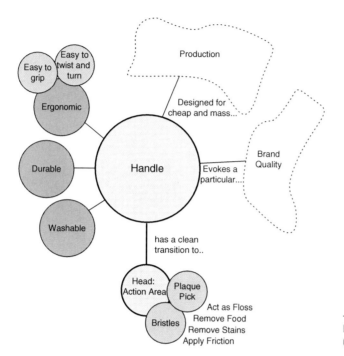

Figure 7.12
Concept map at zoom level 0.

elements have become smaller, so you will need to apply new labels. The map itself takes up less space within its background. Begin to fill in the new background with the elements that exist in the periphery of the map. Adding elements forces you to consider people, technology, entities, ideas, and artifacts that were deemed "out of scope" or only "tangentially related" and to identify the actual relationships between new and original elements. Call this "zoom level –1." Typically, zoom level –1 describes a product line, brand, or business unit. It illustrates the relationships between a product and other offerings from the same company and between a product and other products in the life of a user (see Fig. 7.13).

3. Zoom out and redraw the entire concept map again. The initial elements are now too small to be seen. Add new concepts to the periphery. Label this "zoom level –2," which often describes a company, segment, or sector, and the relationships among business units,

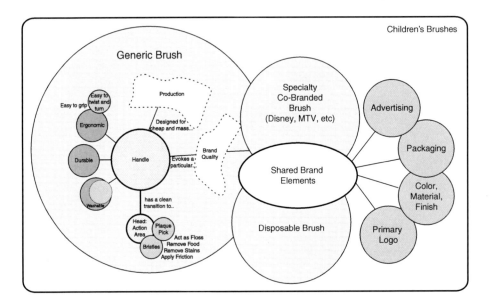

Figure 7.13
Concept map at zoom level −1.

consumers, and entities that affect development and use on a gross scale (see Fig. 7.14).

4. Zoom out one more time, to "zoom level −3." No matter where you started, you now should be thinking about business strategy, societal contexts, and broad issues that affect your product, system, or service. Zoom level −3 examines the world, competition, market, and global issues concerning development, production, distribution, equity, and equality (see Fig. 7.15).

5. Now return to your starting point, zoom level 0, and begin the whole process again, but this time zoom *in*. Focus on a particular feature or function on the concept map that relates to your specific problem. Redraw the concept map and label it "zoom level +1." As you get closer to the material, you view new data about it, so identify new noun and verb relationships (see Fig. 7.16).

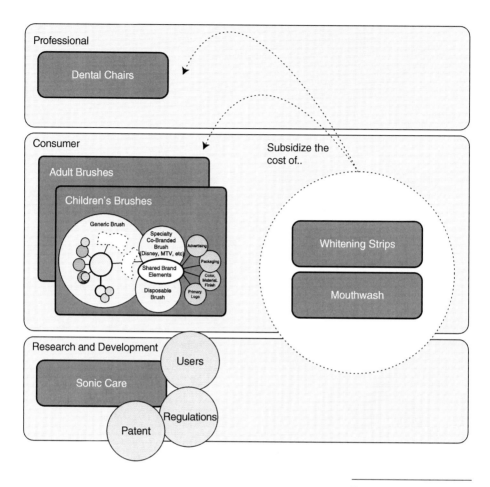

Professional

Dental Chairs ▶

Consumer ▶ Subsidize the cost of..

Adult Brushes

Children's Brushes

Generic Brush

Specialty Co-Branded Brush (Disney, MTV, etc)

Advertising

Packaging

Shared Brand Elements

Disposable Brush

Primary Logo

Color, Material, Finish

Whitening Strips

Mouthwash

Research and Development

Users

Sonic Care

Regulations

Patent

Figure 7.14
Concept map at zoom level −2.

6. When you zoom in again, to "zoom level +2," and redraw the map, you can look at elements *within* a particular feature. In a software product, this might be specific widgets or interface controls that allow a user to interact with the system (see Fig. 7.17).

7. Finally, "zoom level +3" brings you close to a single element of the product. Again, in the context of software, having zoomed in on a dropdown widget, you are now actually considering the list of attributes within it (see Fig. 7.18).

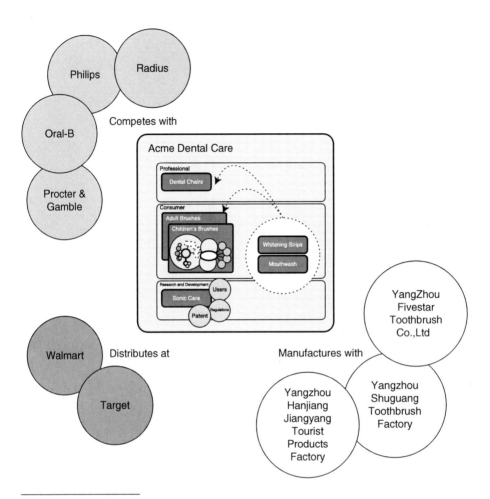

Figure 7.15
Concept map at zoom level −3.

Consider how the seven levels of zoom allow you to traverse the entire concept hierarchy of the design space (see Fig. 7.19).

At each of these levels, you will generate new knowledge about the experience framework being built. All levels of detail will ultimately lead to the "user experience," so this method of synthesis allows you to acknowledge each level individually yet retain a sense of the whole. In this way, you can generate the substance of thought that will drive creative decisions during the design phase of the project.

Exposing the Magic of Design

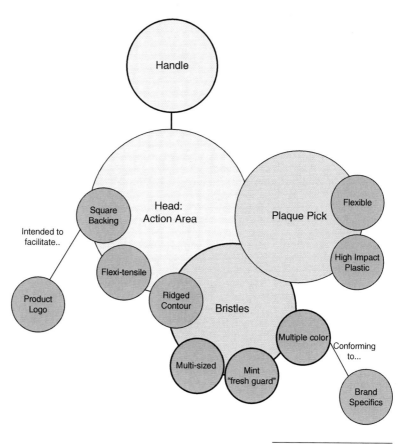

Figure 7.16
Concept map at zoom level +1.

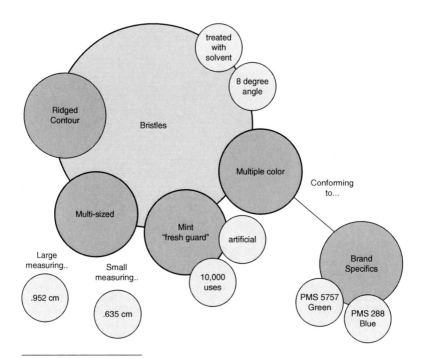

Figure 7.17
Concept map at zoom level +2.

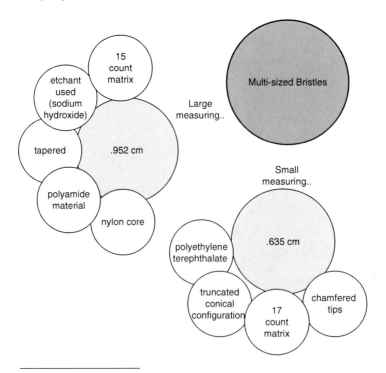

Figure 7.18
Concept map at zoom level +3.

Exposing the Magic of Design

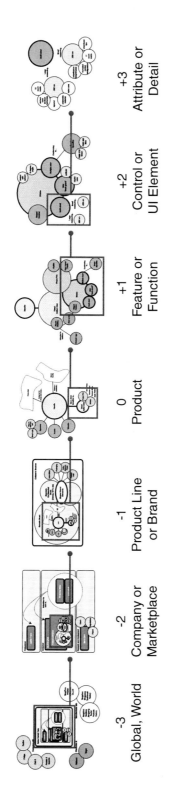

-3
Global, World

-2
Company or
Marketplace

-1
Product Line
or Brand

0
Product

+1
Feature or
Function

+2
Control or
UI Element

+3
Attribute or
Detail

Figure 7.19
Display of all seven of the concept map zoom levels.

For Example:
Breakpoint Diagrams and Other Tools
for Transitions

Contributed by Paul Gould, MAYA Design,
Pittsburgh, Pennsylvania

Diagrams as Strategic Tools for Stakeholder Collaboration

Just about every design project faces three perils:

- *Presumptive misdirection.* Clients who have a lot of information about their customers assume that they have the right information. This is common when they gather feedback largely through surveys or focus groups, or when someone with a strong personality drives the design in a direction they believe to be right.

- *Crippling politics.* With large organizations, multiple divisions and teams, or competing agendas, processes get bogged down and the design ends up favoring the interests of the group that wins out.

- *Tunnel vision.* When people work in organizational silos they tend to address problems in an isolated fashion. Any individual aspect of the final design might be fine, but the overall experience can suffer from incoherence.

To eliminate or bypass these problems, we frequently use diagrams as a lever.

Our clients produce a lot of diagrams, too. These are often standard pie and bar charts that simply report statistics, or dense diagrams that map all the capillaries of a complex system. They tend to disappear after a cameo in a report or presentation.

By contrast, our diagrams are the start of something, not the end. They become collaborative storytelling tools that get stakeholders (makers, funders/buyers, and users) on the same page with a shared understanding and a common purpose.

Our "Breakpoint Diagram" is one such tool. It became a foundation piece and rallying point in a project that awoke a sleeping giant.

The Carnegie Library. The problem: As the primary currency of information has changed from paper to bits, librarians have continued to organize information and connect people to it. But with starvation budgets and competition in the information business, libraries began to suffer from stagnation, entropy, and inbreeding. They assumed a defensive posture, and librarians sometimes seemed more like guardians of information (see Fig. 7.20).

Although librarians perceived themselves as helpful providers of information, users didn't always find them approachable or the information navigable. Libraries ". . . didn't really take into account the experience of people in the building, what they were trying to achieve, why they were there," said Herb Elish, then Carnegie Library's new director. With this in mind, the Library's leaders developed a strategic plan to regain relevance and attract new users, which included understanding library users and their experiences.

Understanding Users. This was strange territory for librarians. They, and many other people who are steeped in an environment or process, think that what they know is just as

Figure 7.20
Example of a reference librarian's desk.

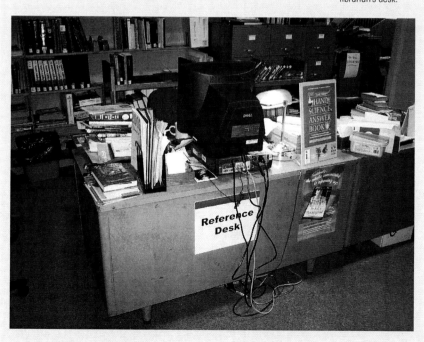

obvious to everyone else. Librarians said, "The experience of users? Isn't that straightforward? Look up an item, find its location, retrieve it, and check it out."

We knew exactly how many questions librarians answered by phone versus in person. We also knew how users had responded when the Library had asked what they want, which included "open longer hours" or "more new materials." But we needed to know their goals and motivations, and we needed to gain insights into how people actually *used* the Library. We began by using ethnographic research methods that included shadowing, interviews, contextual inquiry, and usability analysis.

Where Did Breakpoint Analysis Fit in Our Design Process?
The following describes a high-level flow of the process used in this project:

1. Kickoff (goals, themes, strategy, what is already known)

2. Ethnographic research (understanding users)

    • Shadowing

    • Interviews

    • Contextual inquiry

    • Usability analysis

3. Synthesis (communicating findings and directing design)

    • Personas

    • Breakpoint analysis

    • Affinity clustering

    • Information architecture

    • Lexicon

    • Issue/value analysis

4. Tiger teams (participatory design with clients and users)

5. Prototyping (iterative exploration and refinements)

6. Usability testing (is the design getting better?)

After even just a bit of initial research, our team started forming our first diagram. It described what we called the *components of the library experience* (see Fig. 7.21).

Components of the Library Experience. The components we identified during research represented the fundamental aspects of interacting with the library that were common to every user experience. The most critical insight involves the cluster of things that we labeled "Organizers." Although library interactions took many different forms or channels (telephones, computers, handwritten notes, conversations, signs, and so on), those channels were essentially three types of things:

1. Space: the physical space and structure, and the placement, arrangement, and location of objects in that space

2. Categorizations: classification schemes, labels, lists

3. People: librarians, staff, anyone who has some effect on the experience

Alone and together, these components of the library experience act as organizers of the experience.

This form of expression helps present complex things in a noncontroversial and broadly useful way. In fact, when we introduced this diagram, librarians told us that it summed up the essence of library experience in a way they had often thought about but were never able to express succinctly. (We go big with these documents, working with at least 11x17-inch sheets to encourage recipients to keep them visible as conversation tools. Several librarians displayed this diagram in their offices.)

Breakpoint Diagrams. People who do not use the library were an important target audience. In one case, we assigned one of these nonusers the task of finding a specific book. We asked the person to draw on a map where he went and take a picture (with a disposable camera) every time he encountered

Components of the Library Experience

Users go through organizers to get to materials and activities

We analyzed the users' present-day experience with the library through input sessions with stakeholders, shadowing of users and librarians, and the development of archetypical personas and scenarios. Identified here are the four major components of the library experience.

USERS

... go through ▶ ## ORGANIZERS

Users are at the heart of the library experience. Although the library strives to cater to everyone's needs, we find it useful to focus discussion on a few realistically drawn archetype users. These personas help us define requirements and test design ideas. We will use personas tagged with a star as examples in this document.

The library experience centers on finding answers. To facilitate this, many organizational structures and systems have evolved over the years. These organizers form the primary interface between users and library materials:

▲ **Non-user** ★
Naomi: Seeks a bestseller

Searcher ★
Nicolas: Wants to publish a short story

Learner
Craig: Interested in classes

Media guy
Geoff: Wants entertainment

Explorer ★
Alicia: Wants to attend an event

Teen ★
Frankie: Wants to hang out

Parent
Ellen: Seeks interaction with other moms

Senior
Holly: Seeks social interaction

Shelter seeker
▼ Jack: Wants a safe place

FILTER BY
Subject
Audience (All, Adults, Teens. Kids)
Language
Resources
(Catalog, Deep Web, Internet)

SORT BY
Title
Date/Time
Availability
Location
Popularity
Recommended

 Space
The physical space and structure of the library is the first organizer that most users encounter. The placement of stacks, signage, and even the shapes of desks contribute to the way space organizes library assets and affects interactions with users.

 Categorizations
The library maintains a variety of catalog schemes including the Library of Congress system, the Dewey Decimal System, genre classifications, and even ad-hoc lists. Many of these are accessible electronically.

 People
People are the most flexible organizers, whether it's a librarian giving expert assistance about an abstruse research topic or one's best friend recommending favorite mystery novels.

Figure 7.21
A high-level view of what happens during the experience of most library users.

PROPRIETARY | 01

21 January 2003

Contact: Mickey McManus, 412-488-2900

MAYA Document no. LIBRARY-03003

Final Analysis and Information Architecture

... to get to ▶ **MATERIALS/ACTIVITIES**   ... in order to ▶ **USE / PARTICIPATE**

One of the primary attractions of the library is its free access to diverse, desirable materials and activities.

The most obvious way that users participate in the library is by borrowing an item, taking it home, enjoying it, and finally returning it.

Types of Materials

PHYSICAL COLLECTION
Books
Books on CD, Tape
CD-ROMs
DVDs
Large-print books
Music CDs, LPs, Scores
Videos

DIGITAL COLLECTION
eBooks
Articles/Titles
Government documents
Patents
Reference databases

Types of Activities

SCHEDULED OR AD-HOC
Baby lap sit
Pre-school/Toddler story time
Book discussions/sales
Poetry readings
Art exhibitions
Classes and presentations
Community meetings
Computer classes
Exhibits
Family programs
Films/videos
Lectures/speakers
Music
Play game
Kill time
Be seen
Rendezvous
Private place
Escape reality

User Actions: Materials

Reserve, Check-out, Read/Use, Purchase, Return, Renew, Request more like this, Recommend, Review/Comment, Author

User Actions: Activities

Reserve, Attend/Participate, Request more like this, Recommend, Review/Comment, Create

Users should not simply interact with a material or attend an event, but find that the library sparks additional interests and opens new doors. The library should naturally foster new connections. Ideally, users will benefit not only from these connections, but also help add to them through feedback to the library.

something that blocked him from getting closer to his goal. For example, the computer-based catalog turned out to be unforgiving (e.g., "The Girl With *The* Pearl Earring" returned no expected results and "The Girl With *A* Pearl Earring" did). The participant was well educated and the task was not difficult, but the resulting map of his effort to locate the book looked like spaghetti. This person bounced from one place to another using one organizer after another without ever finding the book. The photos were revealing, too (see Fig. 7.22). Dismayed librarians said, "That's how it is? That's how we look?"

Better Together. Delivering such bad news can be most enlightening, powerful, and tactful when it comes in the form of video, audio, or photographic evidence of real customer experiences, or unbiased usability-test results. Most clients are grateful to discover opportunities to improve and—especially in the case of something new and unreleased—help them to reduce their risk.

Figure 7.22
Photos and notes from the experience of a library nonuser in pursuit of a specific book include many of the roadblocks experienced.

Exposing the Magic of Design

Another way to prepare clients to accept the results is to involve clients directly in the work where possible and appropriate, which makes it more of a codiscovery process. So librarians who at first felt defensive (or at least under the microscope) did start to see things through the eyes of their customers, partly because they saw them firsthand with us. They made major changes in how they worked with customers and now report much higher levels of satisfaction with their work.

But we still needed a way to generalize our findings in a way that kept the experiences at the forefront without getting bogged down in minutiae. We combined our new understanding of organizers, major use phases, users with specific goals and motivations, and documentation of actual user-experience roadblocks or "breakpoints" into a breakpoint diagram (see Fig. 7.23).

As mapped against real experiences from our research, every time that one type of user encountered a breakpoint that prevented moving closer to a goal, we marked it with a big red X. We referred to such experiences as "falling off a cliff" because even when it did not end all forward motion, it plunged users into a different and jarring context (see Fig. 7.24).

Although it is tempting to solve every individual breakpoint as it comes up, doing so would miss out on solving larger issues that can address multiple breakpoints at once. Looking at experience and breakpoint patterns across a series of diagrams for different types of users helped us to make design recommendations beyond the library buildings, such as the following:

- Expanding the notion of usability to encompass a place as an interface and an experience as a product

- Calling attention to unnecessary silos or stovepipes, including administration, information, technology, facilities, organization schemes, policies, terminology, and so on

- Bridging gaps between products and systems

Non-user
Naomi

Primary goal
Wants bestseller

Secondary goals
Check out CDs or DVDs
Get answers about health
or family history

Description
Naomi is 39 and married with two children. She works full time as the Director of Sales for PhilTek Systems in Oakland. She's either traveling or at the office late into the night. She lives with her family in a renovated loft on the South Side. She often orders books and DVDs online using Amazon or Netflix, which are conveniently accessible when she's on the go.

Library experience
Naomi has never been to the library as an adult and thinks of it only as a musty institution from her childhood. She thinks it is unlikely that they have anything "fresh."

ORGANIZERS

Sees an ad for a new Stephen King novel "available for free at your library." Decides to stop at the nearby CLP-Main library after work.

Once at the library, she wonders, "Should I ask a librarian? Should I look it up in the catalog? Where is the catalog? I see some shelves with a sign that says 'popular fiction.' Should I check there?"

Lost opportunity to train the user. How can the library anticipate user goals?

Looks through the popular fiction area, but can't find the book.

Applies for and receives a library card from the registration desk.

Needs to get her car before parking meter runs out. Has no time to return to the librarian to complete the request for the book. Leaves building.

Possible permanent loss of a user. Does the library lose connections with customers when they step out of the building?

Librarian looks it up for her. Determines that it is available at branch library (close to her workplace). Librarian suggests reserving the book.

Wants to request the book. She can't do so without a library card, so needs to apply for one.

Lost opportunity to train the user. User has stated her needs and intentions. How can the library save this state? Can the user get the card in this step?

How can the library take advantage of the previous state? Can the user reserve an item at the same time as getting a card? Are there lost marketing opportunities to hook or speak to first-time customers?

Doesn't really want to ask, she likes figuring things out herself, but finally finds a librarian and asks for help.

Figure 7.23
The breakpoint diagram format communicated details about complex and diverse experiences of key types of library users over major phases of use and across multiple organizers of experience—an accessible way to identify problem areas. It helped clients and librarians to understand the user experience and helped designer to feed designs aimed at eliminating breakpoints and smoothing out transitions among organizers.

Frequency of visit: Never
Help-seeking: Negative
Gender: Female
Age: 20s–40s

MATERIALS

USE

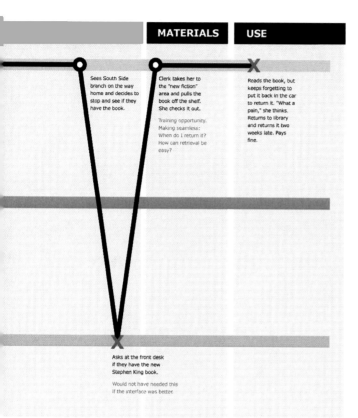

Sees South Side branch on the way home and decides to stop and see if they have the book.

Clerk takes her to the "new fiction" area and pulls the book off the shelf. She checks it out.

Training opportunity. Making seamless: When do I return it? How can retrieval be easy?

Reads the book, but keeps forgetting to put it back in the car to return it. "What a pain," she thinks. Returns to library and returns it two weeks late. Pays fine.

Asks at the front desk if they have the new Stephen King book.

Would not have needed this if the interface was better.

PROPRIETARY | 21 January 2003 | Contact: Mickey McManus, 412-488-2900

MAYA Document no. LIBRARY-03003

Final Analysis and Information Architecture

Figure 7.24
"Falling off a cliff."

About the Author.   As a designer at MAYA Design in Pittsburgh, Pennsylvania, Paul Gould works with clients in diverse domains—health care, energy, finance, defense, consumer products, and public organizations—to make complex information and interactions clear by first understanding human motivations, capabilities, needs, and contexts of use. He has taught human-centered design methods such as ethnographic research, information architecture, and prototyping in corporate settings and in public schools. Paul also speaks at conferences and workshops about topics related to the intersection of design, human experience, and strategic innovation.

Method: Forced Temporal Zoom ("Customer Journey Mapping")

As shown earlier, you can shift semantic perspectives to see things differently. Similarly, you can shift temporal perspectives to understand how a product, service, or system lives or grows with a user. You can look back in time to synthesize, for example, preproduction or supply-chain requirements of physical goods. Or you can look forward in time to understand how a product might be discarded and how to design for disassembly and cradle-to-cradle reuse.

Other names for temporal zoom are "customer journey map" or "life-cycle experience analysis." These names all reference the time-based nature of looking at the extended interactions with a product, system, or service. As an example, consider again Texas Telephone's service configuration problem described earlier. Typically, emphasis will be placed on the time of primary use, or in this example, when someone buys service. However, you can look at both sides of time, understanding first what it was like to research service plans before being a customer, and second, what it will be like to change or cancel service in the future. In both cases, time is shifted, which affords a new set of "what if" questions. By synthesizing these during product development, you can begin to design a cohesive and scalable framework for effective and engaging experiences.

How to Apply This Method

Like a semantic zoom, a temporal zoom starts with a concept map. Again, you will change perspective, but rather than changing the scale of a problem, consider it in the context of a timeline.

1. Sketch a concept map, labeled, as before, "zoom level 0." In the temporal zoom, this level describes a single product, system, or service in its idealized, normal, everyday usage. This artifact does not actually exist, so even the initial zoom level produces generative knowledge. This first step, which might take several iterations, is typically created on a whiteboard by a team of designers (see Fig. 7.25).

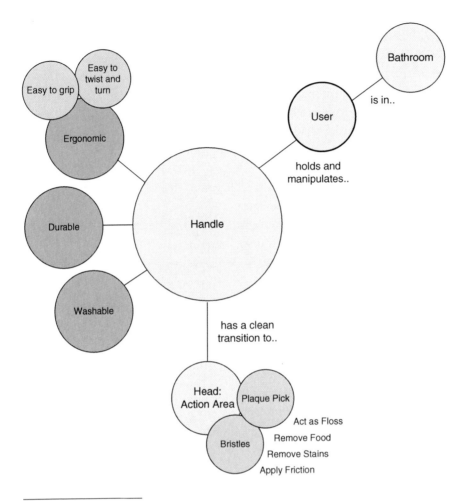

Figure 7.25
Concept map with temporal
zoom at level 0.

2. Now, zoom backwards in time to the *first time* the user
   used the product being described. It may have been in
   a training scenario, a friend may have introduced the
   product, or it may have been in a marketing context.
   Whatever the experience, redraw the concept map to
   emphasize this first use. Label this zoom level –1.
   Emphasis at this stage might be on the out-of-box
   experience, a trial use in a store, or a "viral" training
   session of a friend sharing a design with another friend
   (see Fig. 7.26).

Exposing the Magic of Design

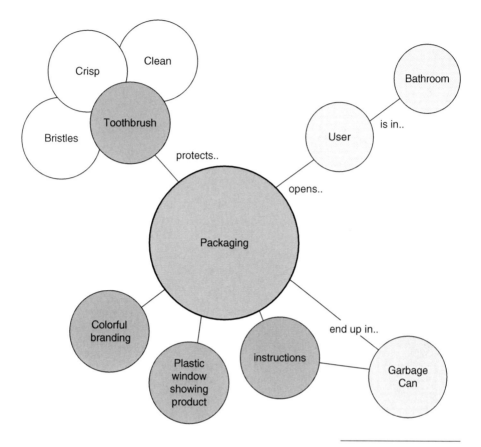

Figure 7.26
Concept map with temporal zoom at level –1.

3. Again, zoom backward again, to zoom level –2, which describes how the user acquired the product *prior to use*. Describe this acquisition phase through the use of a concept map to illustrate, again, the relationship between nouns and verbs. This level usually involves some form of transaction, so describe how a brand is viewed during this phase (see Fig. 7.27).

4. The final backward zoom—level –3—looks at the *preacquisition* experience. This may be how the user learned about the system or item in the first place, or how a friend purchased it as a gift and anticipated that the user would like it. Emphasis is placed on the discovery, both of the brand and the design itself (see Fig. 7.28).

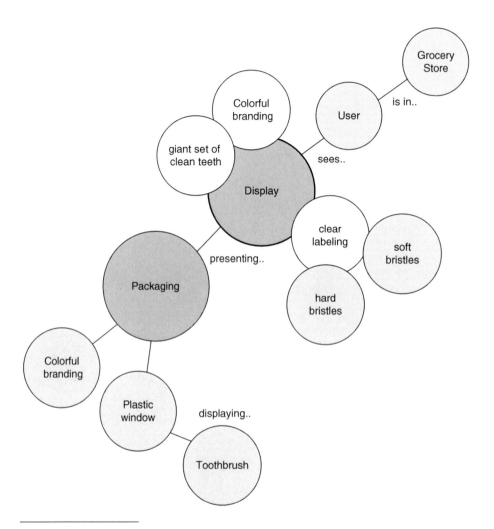

Figure 7.27
Concept map with temporal
zoom at level −2.

5. Returning to your starting point, zoom level 0, you are
   ready to explore what happens to the design in the
   future. First, redraw the concept map to describe how
   the design is viewed by someone who has *developed
   expertise*, someone who has lived with the product
   long enough to learn its nuances, advanced features,
   and functionality (zoom level +1). This step often
   illustrates shortcuts, power functions, and more
   optimized, goal-directed functionality (see Fig. 7.29).

6. Zoom level +2 describes the *failure scenario* where the
   product, system, or service reaches the end of its life.
   This may be caused by technical failure, or it might be

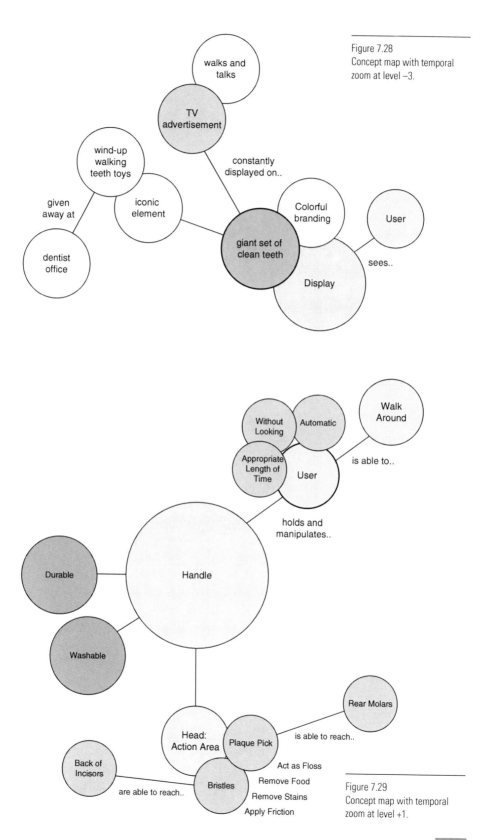

Figure 7.28
Concept map with temporal zoom at level −3.

Figure 7.29
Concept map with temporal zoom at level +1.

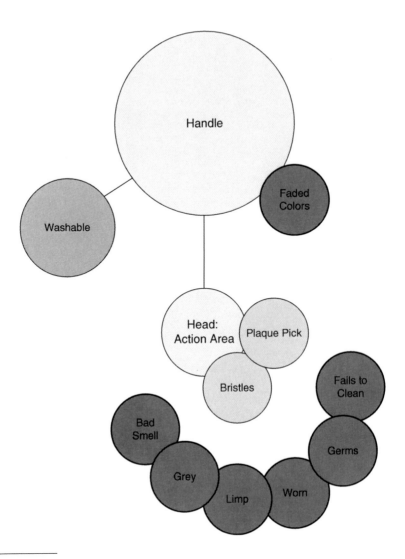

Figure 7.30
Concept map with temporal
zoom at level +2.

due to exhausting a limited quantity of an item. This might even be a more broad failure scenario, described by competitive marketplace trends that lead to the demise of a product, system, or service (see Fig. 7.30).

7. Zoom level +3 describes the *replacement* scenario, where the user replaces the obsolescent, failed, or broken product with a new one. This level raises questions such as these: How does she apply her previous knowledge to the new design? How does she port her content from one digital tool to another? Where does she put the old product? (see Fig. 7.31).

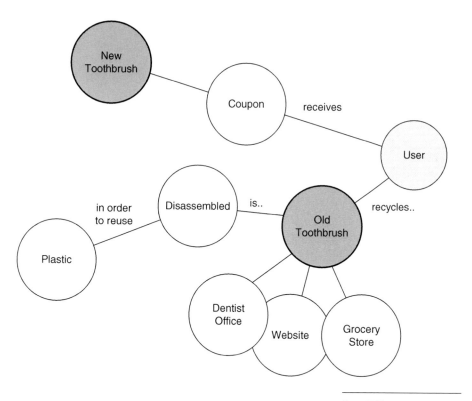

Figure 7.31
Concept map with temporal
zoom at level +3.

Consider how the seven levels of zoom allow you to traverse the product's entire life cycle (see Fig. 7.32).

At each of these levels, you are basically redrawing the
concept map. The focus is on nouns and verbs, and on describing meaningful qualities and elements to consider during
future design phases. When the method is complete, you will
have established a series of maps that have illustrated new
problems, new opportunities, and new ways of thinking about
the entire product, system, or service.

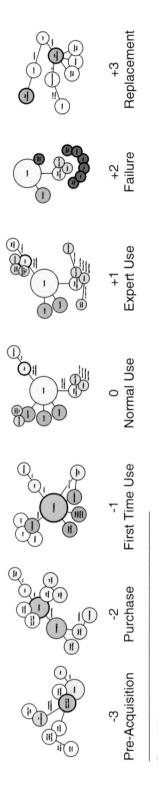

Figure 7.32
The seven levels of temporal zoom and the product's life cycle.

For Example:
The Emotional Touch Points of Shopping

Contributed by Beth Johnson and Gianna Marzilli Ericson,
Design Continuum

Spence Diamonds offered an excellent customer experience—
until the stores became busy. The personalized service was
unable to scale to the large demand, and customers were left
to fend for themselves. They tried to make sense of the variety
of ring settings, eventually became overwhelmed and frus-
trated, and left the store without making a purchase. The
stores could not support the volume of customers they were
attracting.

So the client asked us to design a new, partially self-guided
retail experience to better engage, educate, and empower customers while they waited for a sales associate to become available. Our first step was to create a journey map based on
contextual research, to give everyone involved in the project a
common framework of customer touch points from which to
design.

The contextual research started with "secret shopping" at
the stores to put ourselves in the place of users and to experience noted physical and emotional touch points. We then
talked to customers in the store and to recent buyers to understand their individual interactions. By experiencing the sales
process ourselves, we could map the physical touch points of
the process; talking to real customers helped us understand
the emotional implications of interacting with the brand.

The resulting journey map translated the team's field
observations into a usable design tool that would act as the
backbone for creating an appropriate engagement-ring shopping experience. The journey map was used to visualize the
entire customer experience on one page and find ways to
improve every point of contact with the different types of
customers. The journey map made areas of improvement easy
to identify (see Fig. 7.33).

For instance, the journey map pointed out opportunities
to connect with customers before they engage with a sales
associate. It also helped to guide brainstorms of how to affect
the customer's senses from the moment he pulls into the

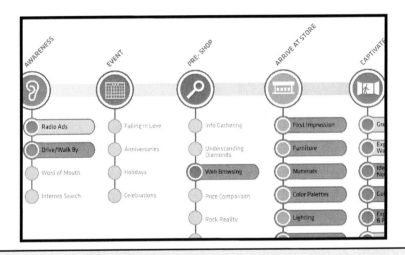

Figure 7.33
Example of a journey map.

parking lot, including a custom mix of love songs and a custom scent for the store. Being able to see the complete customer journey helped to guide every decision from the color palette, to material finishes, to the staff's dress code. More important, the journey map allowed the team to see how all of these varied design choices would interact with each other to create the full experience.

The journey map also helped us to see unmet needs and wishes. It led to the creation of ways to help support the engagement-ring shopper's decision-making process, such as private conversation nooks and a tray that helps shoppers organize their choices. Research and the journey also identified the ring shopper's need for discretion, which inspired the design of receipts with a secret code and a ring sizer that looks like a coaster (see Fig. 7.34).

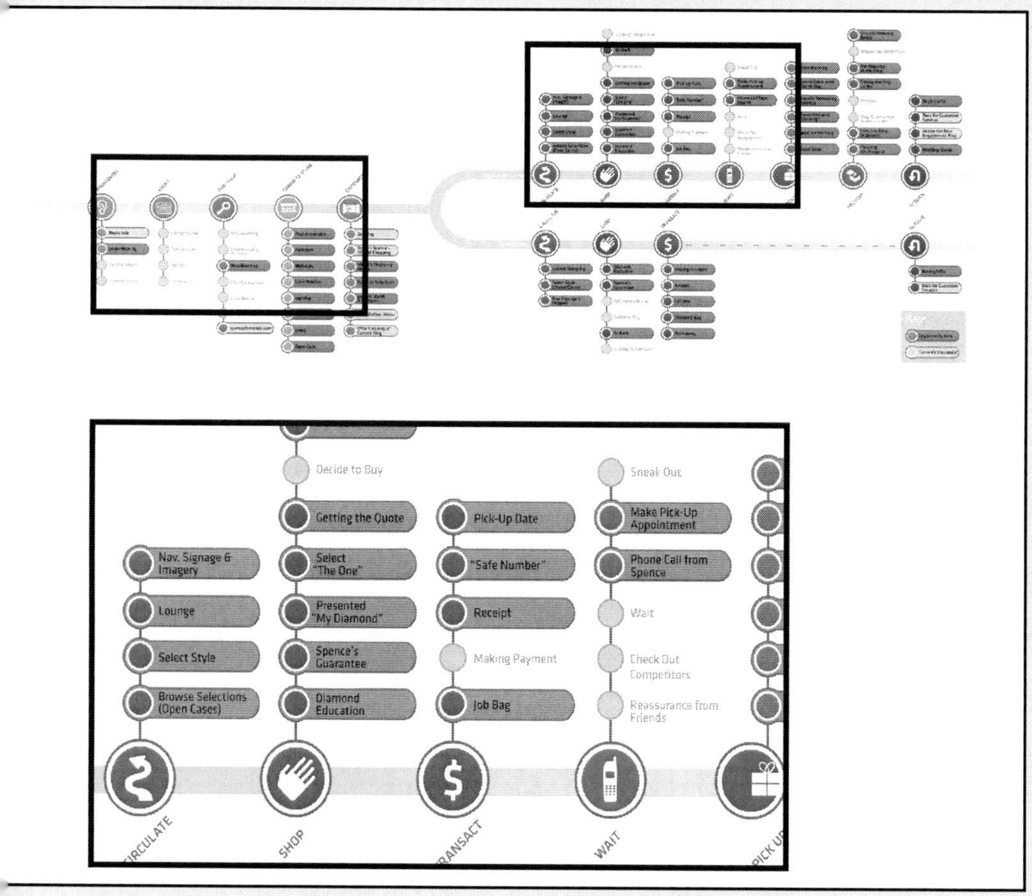
Decide to Buy
Getting the Quote
Select
"The One"
Presented
"My Diamond"
Spence's
Guarantee
Diamond
Education
Nav. Signage &
Imagery
Lounge
Select Style
Browse Selections
(Open Cases)
Pick-Up Date
"Safe Number"
Receipt
Making Payment
Job Bag
Sneak Out
Make Pick-Up
Appointment
Phone Call from
Spence
Wait
Check Out
Competitors
Reassurance from
Friends
SHOP
WAIT

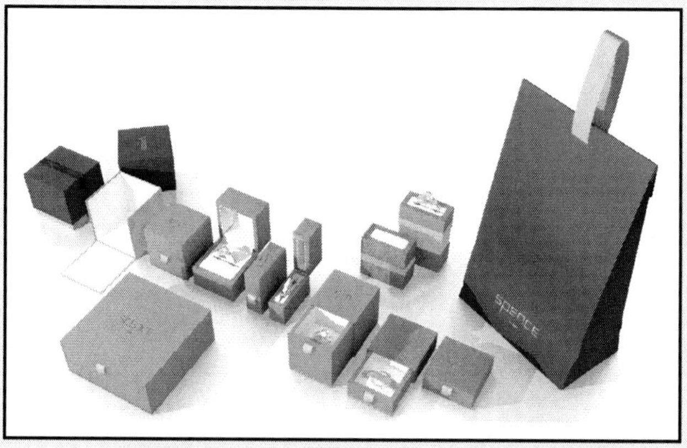

Figure 7.34
The resulting design of these products reflects the usefulness of the journey map in identifying customers' unmet needs.

About the Authors. Beth Johnson is an envisioner at Continuum, where she has focused on business-to-business strategies and created design strategies for Herman Miller, Nestle, and Penske Truck Leasing. Previously, Beth worked in the power tool and baby product industries. She holds a bachelor of fine arts degree in industrial design from the Savannah College of Art and Design and is pursuing her master's degree in business administration.

Gianna Marzilli Ericson brings her background in design and research to comprehensive design programs for Continuum clients across the retail, health care, and service sectors, including Quest Diagnostics, Massachusetts General Hospital, and The Rockefeller Foundation.

Gianna's interests lie in service design, design strategy, health care environments, and communications design. She has prior experience in clinical research and health communications, and she holds degrees from Williams College and The Massachusetts College of Art and Design.

Chapter 8

Methods for Creating Empathy and Insight

Understanding Chasm 3: Empathy and Insight

The third chasm separates knowledge from wisdom. It is perhaps the most difficult chasm to bridge because of the illusive nature of wisdom itself. Many consider that wisdom is achieved through both diverse and deep experiences. For example, a teen living on the streets of New York City may be considered "wise beyond his years," while an author may be praised as "wise" toward the end of her career. Experiences do not have to be long or drawn out; they simply have to be emotionally resonant. The path toward wisdom about design leads through the experience of the beauty, pain, complexity, and harmony of the human condition.

The key to achieving wisdom falls directly within the emotional capacity for empathy. You will need to empathize with the people who will be using, buying, considering, or consuming your design. Empathy is fairly easy to discuss, but it is curiously difficult to *feel*. Most ethnographic tools are used to *understand* context—to uncover details related to workflow, or to learn vocabulary related to a particular group of people or activity. Although this is useful, and particularly important for enhancing usability or adding features and functions to a product, understanding is not synonymous with empathy. To *feel* what it is like to be another individual, you must identify with that person's culture, emotions, and personal style. Tools like personas have been created as a proxy for actual comprehensive emotive immersion, but these tools lack real feelings. They do not capture the depth of sorrow, passion, or hilarity that can be found in actual life and in actual experiences.

That is the real value of empathy. Experiences involve both the pragmatic (activities, goals, and tasks) and also the conceptual and fleeting (such as feelings, irrationality, and culture). Methods that attempt to formalize empathy can help you design for not only utility and practicality but also emotion and behavior—the underpinnings of interaction design and arguably the most important aspects of design in culture.

Ralph Waldo Emerson noted that "a man who seldom rides, needs only to get into a coach and traverse his own town, to turn the street into a puppet-show" (Emerson, 1836). For a designer, the world is the puppet show. It is hilariously entertaining, intellectually instructive, and curiously beautiful when viewed as the subject of design research. Paradoxically, the designer spends an exorbitant amount of his time in the studio or in front of his computer—far from the puppet show. Perhaps, in an effort to move toward a sense of humble wisdom about humanity, the designer needs to live more vibrantly.

Although observing the puppet show of life provides a dramatic amount of insight into humanity, it provides only a sensory glance at the world around us. Intellectual discourse can serve to fill in the gaps, allowing for a bridge between observation and understanding. We can find this discourse in newspapers, magazines, and books. A wealth of knowledge is waiting to be discovered in the written word, and we can use it to attain wisdom. In the same way that a story helps move structured information from information to knowledge, the act of reading can allow for the synthesis of knowledge into a world outlook.

These methods—of moving from knowledge to wisdom—emphasize this vibrancy. The empathy that is established through these methods is an inquisitive feeling of opportunity: to imagine what it is like to be another person, in another situation, with new tools and systems and services. These methods are often the most fulfilling for design teams because they "feel right": The methods act as extensions of characteristics you likely already have.

Method: Reframing

You will always approach creative problem solving in the conceptual context of a "frame," as described earlier. This frame creates a subjective boundary around a problem space, and that boundary becomes a relatively objective constraint by which problem solving becomes easier.

The frame is usually selected without introspection, based on experience, research, and assumptions. Consider the example of being tasked with creating an "innovative new toothbrush." You will probably create (automatically, and without much thought) a frame like this:

> *an average person, in his bathroom, using a physical item with small bristles on the end to apply paste to his teeth; that individual will likely produce friction with the physical item, the paste and the teeth in order to remove food particles on the teeth.*

Note that this frame describes a person, a setting, and an action-based goal. It describes a very culturally specific and archetypical example of teeth brushing.

The design method of *reframing* recasts the normal frame in a new, often unexpected perspective. Consider reframing this example from the perspective of someone different than the nondescript "average person." You can purposefully view the problem from the perspective of a dentist, a toothpaste manufacturer, a child, someone from Thailand, someone with no working limbs, or a group of people. The implications for designed artifacts are dramatically shifted each time the problem is reframed, and new design ideas—and design constraints—emerge with each new frame.

Thus, *reframing is a method of shifting semantic perspective to see things in a new way.* The new frame "reembeds" a product, system, or service in a new (and not necessarily logical) context, allowing you to explore associations and hidden links to and from the center of focus.

This method attempts to move from knowledge to wisdom, because it demands that you empathize with a target audience by trying to understand how something would *feel* in a particular situation. Additionally, this method requires that you tell a short but compelling story about a particular new perspective, emphasizing both behavior and empathy. The method assumes that you have enough tacit knowledge and background in the working space to be effective in using empathy. For instance, it is difficult to empathize with a dentist if one has never seen or interacted with a dentist in a meaningful way.

How to Apply This Method

To reframe something, follow these steps:

1. *Identify the "normal" frame.* Unlike in the toothbrush example, that can be difficult. A more realistic example might be the design of a complicated piece of software that is intended to allow for pricing and configuration of parts.

   For the purposes of this method, a design-specific frame can be described as *someone, in an environment, using or considering a particular design embodiment from a particular perspective.* Note the use of "design embodiment," rather than "product." This is because it is possible to reframe things that are not discrete products, such as services, systems, Web sites, and even ideas, philosophies, and assumptions. Also note again that the level of specificity of the perspective, environment, and embodiment are dependent on the design problem being considered. It may be easy to define the frame of a "contained" design problem very specifically, whereas more complicated systems or services problems may require a more robust framing description.

2. *Create blank reframing charts.* You will need three charts to reframe the design opportunity from three new points of view: perspectives, environments, and embodiments. Each chart will look like the one in Table 8.1.

Table 8.1 Example of a Blank Reframing Chart

Reframed from a New [Perspective/ Environment/ Embodiment]	Primary User Goal	Design Implications
# ...		

Exposing the Magic of Design

3. *Reframe.* Through structured or casual brainstorming, the designer will begin to develop new items for the left column of each chart. Depending on the desired level of innovation for the particular design problem, it is often desirable to include "provocations"—ideas that may ultimately prove infeasible but allow for "movement" across patterns (De Bono, 1995).

   There are infinite "answers" to this left column, so this activity can seem both tedious and challenging at once. It is useful to set an arbitrary goal of "number of new frames" and to target more than 100. The first few will be easy, and because they are easy, they will likely be obvious. Less obvious reframes have the potential to offer innovative (although not necessarily useful) design implications. As an example, a toothbrush can be easily considered from the perspective of a dentist, but is it possible to view a toothbrush from the perspective of a dog or a pumpkin, or from outer space?

4. *Extrapolate likely user goals.* As the charts become populated with new frames, the designer will begin to fill in the primary user goal for all items in all charts. She will paint a picture of a credible story, judging responses and adding criticism as appropriate. As the designer thinks of the new frame, she tells a short but compelling—and believable—story of how a user's goal might change given the new frame of reference. This need not be long. A single sentence is sometimes sufficient. When reframing toothbrushing from a dog's perspective, you might say that the dog wants to escape, clean its teeth without feeling anything, or feel as if though it is eating a delicious steak.

   Because extrapolating user goals requires empathy, it is a personal and projective activity. That is, a designer needs to think and feel what it is like to be in another situation. It is common to hear things like "If I were a dog, I would . . ." or "If I were a fire hydrant, I would . . ." during this activity.

5. *Extrapolate design implications.* The reframed design
    context will produce new constraints or implications
    or highlight hidden or overlooked existing ones. These
    design implications may take the form of constraints
    or recommendations—"the design must . . ." or "the
    design should . . . ," which will be useful during the
    design phase.

An alternative approach is to draw the design implications,
rather than listing them in words and phrases. This requires a
team member that is adept at quick sketching. The increased
level of fidelity of the output can have a powerful sense of
resonance for the design team, particularly when the team
includes engineers and software developers. Seeing the ideas
come to life can be persuasive, and the visualization process
can act as a bridge over the pragmatism of "what can actually
be done."

During synthesis, a designer can use the reframing method
to shift frames explicitly and fundamentally, changing the
selected features and relationships and actively producing
new design implications and constraints.

Method: Insight Combination

A common goal for designers is to identify interesting ideas
and to come up with a multitude of potential designs, leaving
the refinement for later. Often the motivation of this activity
is to offer a view of what "could be," not necessarily what
"should be." This allows more conservative stakeholders to
envision the future and dream about directions for products
and services. These dreams are fed by insights, which are in
turn fed by design research. This manner of moving from
research to insight to idea is formulaic, relying on a design
team's active and iterative approach often known as "rigorous
design process." However, the process itself rarely lives up to
the name, because a rigorous process implies documentation,
control, purpose, and thought that simply is not present when
most designers tackle problems of innovation.

Insight combination is a method that generates a multitude of new ideas and embeds these ideas firmly in the opportunity area and in the cultural context of your team. This

Exposing the Magic of Design

implies that design ideas will be tied to research findings and to patterns in society and culture.

Insight combination is a method of building on established design patterns to create initial design ideas. Through multiple steps, this method first demands the articulation of individual design insights and then forces a structured and formal pairing of insights with existing patterns. This pairing creates a new design idea that has a strong connection to both established best practices and to problem-specific research data.

A design insight can be thought of as the additive of problem-specific observation ("I saw this") and personal and professional experience ("I know this"). This grounds an insight in both the subjective and general knowledge of the specific practitioner and in the objective data of the design problem itself. From a sensemaking perspective, this embraces the episodic and experiential uniqueness of your memories and pairs it with generally accepted ways of doing things. An insight might manifest itself in the form of a new, subjective design constraint that you add to the problem, or it might come as an underlying philosophy and approach (a set of guiding pillars or themes that guide the creative efforts). Insights might arise through some of the other methods described in this text, such as affinity diagramming, or they might come "as a flash," as your brain offers its own implicit synthesizing of the problem space.

An insight is almost always contextual to the given design problem being solved. As your personal and professional experience is unique to you, so too is the problem space's experience unique to the domain being studied. In the context of designing a mobile phone interface, for instance, you might apply your expertise in a very different manner than when designing a piece of enterprise software; yet in both cases, your design process might appear to be the same, with the steps and methods following a similar sequence and style. This contextualized approach to design is because of the problem-specific information and how it melds with your personal design philosophy and tacit knowledge.

An insight might be: "Users sometimes have to send data to the person they are talking to, and so the phone should allow them to do this easily without hanging up." Or: "Salespeople will often sketch a hardware configuration while talking to a customer, so the software should help them do this." In both cases, the insight is first observatory, then, prescriptive. It is based on observation, but the observation has been filtered and manipulated, and it has changed a great deal.

As a method, insight combination takes these observatory and prescriptive insights and combines them with design patterns. Through this combination, you will be forced to examine and consider each unique insight and pattern. Methodically, you must think about each facet of the design problem that has been deemed useful or important. The method is then divergent, because it actively produces new ideas. Ideas are "moved forward" in a nonlinear fashion, jumping over the expected to arrive at the unexpected.

How to Apply This Method

Insight combination is best performed after contextual, qualitative research has been conducted. The method relies heavily on the presence of both insights and patterns, and while both are easily identifiable in data, both are also time consuming to produce and to capture in a succinct manner. Research must be analyzed to produce insights. Patterns can be identified only through introspection and reflection.

Insight development is best done in groups and in a casual, distraction-free environment. Because insights come from new ways of looking at data, other methods described in this text, such as affinity diagramming, can be used in tandem with insight combination.

The method of insight combination can be conducted as follows:

1. *Identify insights in the gathered data.* You can begin to identify insights in the data that have been gathered during research by combining an observation (I saw this) with knowledge (I know this). You can then write the insights on yellow note cards. As an example, perhaps you observed someone brushing her teeth and noticed that the individual avoided using the mouthwash that was sitting next to the sink. You might recall your own last visit to the dentist. An insight

could then be developed: that mouthwash has an implicit connection of taste and smell with going to the dentist, which casts the product in a negative light. Of course, this insight could be completely wrong— and that is perfectly acceptable.

2. *Identify design patterns relevant to the core domain.* Recall design patterns that are relevant to the discipline being studied, and write these patterns on blue note cards. Some designers keep pattern libraries, noting trends and repeated design elements that appear in produced artifacts. Others prefer to search for patterns in the context of the problem. A pattern that is loosely related to the toothbrush example might be the trend in consumer goods (kitchen soap, gum, etc.) to introduce new artificial flavors and smells like amaretto and butterscotch. Another pattern is the push toward digital timekeeping devices in children's toothbrushes—ways of helping children keep track of how long they have been brushing.

3. *Perform an insight combination by pairing a design pattern with an insight and looking for affinities.* There is no "method" to this portion of the technique; cards are literally combined at random, by mingling the blue and yellow notes. You might move them around physically and actively reflect on potential combinations. When a combination makes sense and generates a design idea, the idea is written on a green note. Combining the insight (mouthwash has an implicit connection of taste and smell with going to the dentist, which casts the product in a negative light) and the pattern (the trend in consumer goods— kitchen soap, gum—to introduce new artificial flavors and smells like amaretto and butterscotch) yields a new design idea: produce a mouthwash that has a new flavor, one that does not have properties normally associated with the dentist's office.

Many design ideas will be created; but not all of them will be "good ideas." Some will be technically impractical, while others may not have business resonance. During synthesis and particularly while conducting this method, it is best to suspend explicit judgment of ideas and instead focus on producing as many ideas as possible. Later evaluation phases can determine which ideas are best, or most likely to be produced. The focus of this method is on a large quantity of new ideas.

Conclusion

This text has presented a theory of synthesis: a combination of cognitive psychology theory and social psychology knowledge that results in a way of combining data in a well-formed manner. The text has also presented the spark of synthesis: the moment of creative energy that shifts the existing to the potential, and that illustrates the ability for designers to produce new, novel, and emotionally charged design solutions to complex problems. Additionally, the text has illustrated the need for synthesis in the context of business, because synthesis offers a rigorous path toward innovative ideas and a manner of approaching these complex problems in a methodical manner.

By using the methods presented in this text, designers should be able to better make sense of complicated situations and approach complex problems with a new and thorough approach. You will be able to bring rigor to what has traditionally been a flexible and haphazard process. And you will be able to rationalize and better substantiate design decisions, because you will have a clear understanding of the path you used to arrive at an idea—and you will be able to articulate that path succinctly.

Some of the methods are clearly intended for the early stages of synthesis, where you are attempting to understand the problem space and make sense of the data you have gathered. At that stage, a focus on data organization, pruning, and judging is critical in order to make sense of a chaotic and seemingly overwhelming amount of data.

Other methods are used to contextualize your design problem in its larger setting by considering alternative perspectives, new viewpoints, and by embracing the abductive leaps that are required in synthesis. At this stage, a focus on innovation is supported by cultural aspects of play, flow, and storytelling.

The methods that have been presented are intended as starting points, upon which you can build your own tools and techniques. The specifics of the method are incidental. What is important is a rigorous approach to synthesis, one that is made explicit through repeatable techniques and one that can be both documented and rationalized.

Design synthesis is a way of thinking about complicated, multifaceted problems with a repeatable degree of success. Design synthesis methods can be applied in the context of business, with the goal of producing new and compelling products and services. The principles and methods presented in this text are teachable, repeatable, and understandable. They are creative activities that actively generate intellectual value, and they are activities that are unique to the discipline of design. Most important, when applied and formalized, these activities are billable and immensely useful in the development of new, novel, useful, and appropriate designs.

Glossary

Abductive argument. The argument from best explanation, depending on circumstances and experience.

Affinity diagramming. A method of bottom-up data organization intended to define categories based on likeness of data.

Concept map. A graphical method for organizing and representing knowledge.

Constraint. A boundary condition that defines what is considered a successful solution to a design problem.

Deductive argument. The output is guaranteed to be true, if the premise is true.

Design research. A human-centered approach to understanding culture, by observing and then celebrating the unique and peculiar aspects of humanity.

DIKW. The path from *d*ata, to *i*nformation, to *k*nowledge, to *w*isdom.

Divergent thinking. Part of the design process that refers to rapid ideation, producing as many ideas as possible.

Ethnography. A form of qualitative research that requires immersion in the natural contexts upon which work, play, or culture are experienced.

Flow. An optimal experience achieved during creativity that is an automatic, effortless, yet highly focused state of consciousness.

Flow diagramming. A method for visualizing the path through a system.

Forced semantic zoom (ecosystem mapping). A method for understanding—and visualizing—the larger system of people, products, services, and environments within which a particular product exists.

Forced temporal zoom (customer journey mapping). A method for understanding—and visualizing—the situational aspects within which a particular design element exists.

Frame. A perspective or viewpoint.

Inductive argument. Gives good evidence that a conclusion is true.

Innovation. Something new that is successful in the marketplace.

Insight combination. A method of building on established design patterns to create initial design ideas.

Marketing research. A human-centered approach to predicting behavior, usually related to purchasing, through observation and statistical modeling.

Mental model. A cognitive representation and simplification of a complex system.

Reframing. A method of shifting semantic perspective to see things in a new way.

Pattern. A design paradigm, illustrating habitability—something that is beginning to be found in more than one product, system, or service.

Satisfice. A goal to meet objectives but not necessarily to exceed them.

Sensemaking. A process that describes the patterns one sees, the constraints one applies, and the mental models one forms about his or her design problem.

Synthesis. An abductive sensemaking process of manipulating, organizing, pruning, and framing data in an effort to produce information and knowledge.

Visualization. The act of externalizing ideas in a visual format.

Works Cited

Alexander, C. (1964). *Notes on the synthesis of form.* Cambridge, MA: Harvard University Press.

Buchanan, R. (1996). Wicked problems in design thinking. In V. Margolin & R. Buchanan (Eds.), *The idea of design* (p. 9). Cambridge, MA: MIT Press.

Consumer electronics returns are a $13.8 billion problem in the U.S. likely to get worse. (2008). *TechPulse360.com.* Retrieved from http://techpulse360.com/2008/12/16/consumer-electronics-returns-are-a-138-billion-problem-in-the-us-likely-to-get-worse/

Coyne, R. (1988). *Logic models of design.* London, England: Pitman.

Craik, K. (1967). *The nature of explanation.* Cambridge, England: Cambridge University Press.

Csikszentmihalyi, M. (1997). *Creativity: Flow and the psychology of discovery and invention.* New York, NY: Harper Perennial.

De Bono, E. (1995). Serious creativity. *The Journal for Quality and Participation, 18,* 12.

De Bono, E. (1999). *Six thinking hats.* Boston, MA: Back Bay Books.

Dewey, J. (1997). *Experience and education.* New York, NY: Free Press.

Dubberly, H. (2009, May/June). Models of models. *Interactions, 16*(3), 54–60.

Eames, R., & Eames, C. (Writers/Directors). (1978). *Powers of ten* [Film]. New York, NY: IBM.

Emerson, R. W. (1836). *Nature.*

Fallman, D. (2003). Design-oriented human-computer interaction. In *Proceedings of the SIGCHI conference on Human factors in computing systems* (pp. 225–232). Ft. Lauderdale, FL: ACM.

Gedenryd, H. (1998). *How designers work - Making sense of authentic cognitive activities.* London, England: Lund.

GfK Custom Research North America. http://www.gfkamer-ica.com.

Gladwell, M. (2007). *Blink: The power of thinking without thinking.* Boston, MA: Back Bay Books.

Holtzblatt, K., & Beyer, H. (1997). *Contextual design: A customer-centered approach to systems sesigns.* Morgan Kaufmann.

Johnson-Laird, P. (2005). The shape of problems. In V. Girotto (Ed.), *The shape of reason: Essays in honour of Paolo Legrenzi* (pp. 3–26). Psychology Press.

Johnson-Laird, P. (2006). Mental models, sentential reasoning, and illusory inferences. In C. Held, G. Vosgerau, & M. Knauff (Eds.), *Mental models and the mind* (p. 138). New York, NY: Elsevier.

Johnson-Laird, P. (2009). *How we reason.* New York, NY: Oxford University Press.

Klein, G., Moon, B., & Hoffman, R. (2006a). Making sense of sensemaking 1: Alternative perspectives. *Intelligent Systems, 21*(4), 71.

Klein, G., Moon, B., & Hoffman, R. (2006b). Making sense of sensemaking 2: A macrocognitive model. *Intelligent Systems, 21*(5), 91.

Kolko, J. (2007a). Information architecture and design strategy: The importance of synthesis during the process of design. *IDSA 2007 Educational Conference Proceedings.* San Francisco, CA: IDSA.

Kolko, J. (2007b). Information architecture: Synthesis techniques for the muddy middle of the design process. *23rd International Conference on the Beginning Design Student Proceedings.* Savannah, GA:

Lehrer, J. (2009). *How we decide.* New York, NY: Houghton Mifflin Co.

Martin, R. (2009). *The opposable mind.* Cambridge, MA: Harvard Business Press.

Neuhart, J., & Neuhart, M. (1989). *Eames design. The work of the office of Charles and Ray Eames.* New York, NY: Harry N. Abrams.

Novak, J. D., & Cañas, A. J. (2006). *The theory underlying concept maps and how to construct them* (Technical Report

IHMC CmapTools). Pensacola, FL: Florida Institute for Human and Machine Cognition.

Osborn, A. (1963). *Applied imagination; Principles and procedures of creative problem-solving.* New York, NY: Scribner.

Peirce, C. S. (1998a). On the logic of drawing history from ancient documents. In Peirce Edition Project (Ed.), *The essential Peirce: Selected philosophical writings, 1893-1913* (p. 95). Bloomington, IN: Indiana University Press.

Peirce, C. S. (1998b). Pragmatism as the logic of abduction. In Peirce Edition Project (Ed.), *The essential Peirce: Selected philosophical writings, 1893-1913* (p. 227). Bloomington, IN: Indiana University Press.

Shedroff, N. (2000). An overview to understanding. In R. S. Wurman (Ed.), *Information anxiety 2* (p. 27). Indianapolis, IN: Que.

Simon, H. (1973). The structure of ill-structured problems. *Artificial Intelligence, 4,* 181–201.

SIMPLICITY Consortium. Retrieved from http://simplicity.media.mit.edu/

Takeda, H., Tsumaya, A., & Tomiyama, T. (2007). *Integration of knowledge in synthesis process.* Retrieved from http://citeseerx.ist.psu.edu/viewdoc/download?doi=10.1.1.46.6407&rep=rep1&type=pdf

Veen, J. (2000). *The art and science of web design.* London, England: New Riders Press.

Vogel, C., Cagen, J., & Boatwright, P. (2005). *The design of things to come: How ordinary people create extraordinary products.* Philadelphia, PA: Wharton School Publishing.

Wicks, J. (2006, May 17-18). *Weaving design into Motorola's fabric.* Presentation at the Institute of Design: Strategy Conference, Chicago, IL.

Index

later stage task flow diagram in, 99*f*

overview, 94

task flows and, 94–95

user prototype adjustments made in, 98–100, 99*f*

user prototype evaluation in, 97–98

ideas

creative association and, 56

design process regarding, 42

results production concerning, 42–43

visualization and, 56

ideation stage, 38

ill structured problem, 6–8, 10

immediate feedback, 51

immersion perspective

current state/realistic, 34, 35

current state/subjective, 34, 35–36

design research and, 34–36, 36*f*

opportunity/potential, 34–35, 36

implications, 73–74, 164

incomplete information, 9

inductive argument, 24, 172

information. *See also* meaning making techniques

DIKW system and, 59–60, 60*f*, 61*t*

flow model and, 87

implications and, 73–74

insights and, 73

as meaningful data, 59

what it means phase and, 72–73

informed hunch, 8

innovation, 172

design research link, 37–40, 40*f*

design synthesis and, 38–40, 40*f*

expectations, xvi

immersion perspectives and, 34–36, 36*f*

what it is, 37

insight, 73

combination method, 164–68

understanding chasm of, 159–60

insight combination, 172

affinities and, 167

context and, 165

how to apply, 166–67

judgment suspension regarding, 167–68

method, 164–68

observation/prescription and, 165–66

patterns and, 166, 167

rigorous design process and, 164

what it is, 164–65

when to use, 61*t*

intuition, 20, 21, 66

Johnson, Beth

about, 158

"The Emotional Touch Point of Shopping" by, 155–58

judgment

data and, 66

making, 8–9

suspension of, 167–68

knowledge

concept mapping and, 106–7

defined, 59

DIKW system and, 59–60, 60*f*, 61*t*

funnel, 27

reframing and, 161–62

tacit, 8

legal moves, 5

library example, Carnegie

breakpoint diagrams and, 139, 142–43, 142*f*, 144*f*

components of library experience in, 139, 140*f*

design perils introducing, 136

design process, 138–39

"falling off a cliff" and, 143, 146*f*

problem in, 137, 137*f*

recommendations from, 143

understanding users in, 137–38

Making Meaning (Bruner), 68–69

manipulation of time, 94

marketing research, 172

characteristics of, 32*t*

design research *vs.*, 31–33, 32*t*

Martin, Roger, 27, 58

meaning

data and, 63–100

information and, 59

story and, 68–69

meaning making techniques.

See also affinity diagramming; flow diagramming

chasm and, 60, 60*f*

conclusions about, 74

enhancing data through intuitive leaps, 66

theory, design synthesis, xv–xvi.
  See also experience; frame;
  patterns; sensemaking
  benefits of studying, xvi–xvii, 3
  constraints, understanding/
    breaking, and, 9–10
  hierarchical problem solving
    and, 4–5, 5f
  ill structured problem and,
    6–8, 10
  informed hunch and, 8
  judgment making and, 8–9
  partial information and, 9
  problem solving and, 4–10
  rationale for studying, 3–4
  well structured problem and,
    5–6
thermostat example, Hunter fan
  case study, 96–100
  early stage task flow diagram in,
    97, 98f
  final design in, 99f, 100
  of flow diagramming,
    94–100
  later stage task flow diagram in,
    99f
  overview, 94
  task flows and, 94–95
  user prototype adjustments
    made in, 98–100, 99f
  user prototype evaluation in,
    97–98
thinking hats, 49–50

trust building, 110–11
tunnel vision, 136

"Using Concept Maps in Product
  Development" (Dubberly),
  109–24

Veen, Jeff, xi
visualization, 172
  benefits of, 54–56
  bottleneck and, 54
  change over time and, 55
  comparison and, 54–55, 55f
  creative association and, 56
  culture of synthesis and, 54–56
  ideas made concrete via, 56
  spatial relationship and, 55

well structured problem, 5–6
what it means phase, 72–73
Wicked Problems in Design
  (Buchanan), 103–4
Wicks, Jim, xi
wisdom
  DIKW system and, 59–60,
    60f, 61t
  as enlightenment, 59
  path to, 159
  reframing and, 161–62

zooming, 125. *See also* forced
  semantic zoom; forced
  temporal zoom

Printed in the USA/Agawam, MA
February 18, 2015

608903.010